Concise notes on
# MATERIALS SCIENCE AND ENGINEERING

**George O. Rading**
*University of Nairobi*

*Trafford rev. 10/25/2019*

www.trafford.com
North America & international
toll-free: 1 888 232 4444 (USA & Canada)
fax: 812 355 4082

# ABOUT THE AUTHOR

George Odera Rading is currently an Associate Professor at the Department of Mechanical Engineering, University of Nairobi. He obtained a first class honours degree in Mechanical Engineering from the University of Dar es Salaam in 1979. After a brief period during which he worked as an Assistant Engineer with the E. A. Power and Lighting Company (now KPLC), he enrolled for a post-graduate degree program at the University of Nairobi. He was awarded the M.Sc. degree in Mechanical Engineering in 1983. In 1982, he joined the Materials Testing Laboratories at the Kenya Bureau of Standards rising through the ranks to become head of that department in 1986. During the same period, he lectured on a part time basis at both Nairobi and Moi Universities.

Professor Rading joined the full time staff of the Department of Production Engineering at Moi University in 1987 and headed the same Department up to 1990 when he won a Fulbright Scholarship to study for the Ph.D. degree at the University of Alabama, USA. He was awarded the Ph.D. degree in Metallurgical and Materials engineering in 1994. Upon graduation, he was appointed an Assistant Professor of Engineering Science and Mechanics by the same University. He joined the University of Nairobi upon his return from the USA. He was elected to the Kenya National Academy of Sciences in 2004. He has also served a one-year tour of duty as a Visiting Senior Lecturer and Director of Quality Assurance at the Kigali Institute of Science and Technology in Rwanda.

A Registered Professional Engineer, Professor Rading is a member of several Professional Societies and Institutions including the Institution of Engineers of Kenya, American Welding Society and the International Materials Society (ASM).

# CONTENTS

# PREFACE

This book is a compilation and summary of the lectures I have given over the past several years at various Universities in the subject of Materials Science and Engineering. It covers the entire undergraduate syllabus in the subject as specified by most Universities for students studying for the degree course in Mechanical or Production Engineering. Though intended primarily for undergraduate degree students, Higher National Diploma (HND) students may find some parts of the book useful. Practicing Engineers who wish to make a quick reference will also find the book handy.

I was motivated to write this book by two main factors. Firstly, Materials Science and Engineering is a broad and growing subject. While teaching the course, I had particular difficulty recommending a single textbook that covered the entire subject matter to the breadth and depth required by the syllabus. This was further complicated by the fact that the subject matter falls into two distinct categories: the phenomenological (mechanics) and the mechanistic (science). Available textbooks that give excellent treatment to one category of topics inevitably give only a superficial treatment to the other category. For example, most books that give an elaborate account of topics like crystal structure and dislocation theory may dismiss topics like fatigue or mechanics of composite materials in a single paragraph. As someone with a background in both Mechanical Engineering and Materials Engineering, I felt a need to bridge this gap. Secondly, the good textbooks currently available in both the mechanistic and phenomenological aspects of the subject are large and expensive (being largely foreign published) and are more suited for reference. I felt a need for a book that is written in concise form, that covered the entire syllabus, and that students could afford to buy and keep for revision purposes. To this end, the presentation has been kept as short as is possible without losing clarity. A lecturer teaching this subject can then concentrate on expounding the principles behind the theory without wasting class time giving notes. The student on the other hand can spend class time listening and following the development of the theory. To augment this, he/she may consult the reference text that treats a specific topic best. At revision time, "Concise Notes" will provide a handy source of revision notes.

Materials Science and Engineering concerns itself with relating the internal structure of a material to its properties. Moreover, the Engineering part of it seeks to enable one to arrive at a defined set of properties through manipulation of the internal structure. This book has been arranged with this need in mind. Thus, the first three chapters deal with the internal structure of materials that also helps bring out the difference between the three primary classes of materials i.e., metals, polymers and ceramics. The next chapter deals with the properties of materials that are likely to interest the Engineer. Deliberately, emphasis has been placed on mechanical properties considering that the target groups are Mechanical and Production engineering students. The following chapter is on "equilibrium diagrams" that is necessary for the understanding of metallic alloy systems. The next three chapters are dedicated to the first primary class of materials i.e., metals. In the said chapters, the properties of metals and their alloys are looked at in some detail and explained in terms of the nature of their interatomic bonds. Furthermore, methods that may be used to engineer the properties e.g., alloying, heat treatment, and processing are considered for the most important alloy systems (iron, aluminium, copper and titanium based alloys). In the following three chapters, attention is focused on the more specialized material properties i.e., fatigue, creep and fracture. The other two primary materials i.e., polymers and ceramics are then considered. In both cases, the internal structures are considered followed by the properties and finally some brief mention of the major processing methods for each group.

Having studied the primary classes of materials, their "engineered" counterpart, "composite materials" is then considered. Finally, some aspects of surface stability are touched on in the chapter on corrosion and degradation of materials.

Several of the chapters are treated in this book in an introductory manner only. This again was deliberately done. As stated earlier, the subject is broad and new subject matter is being added continuously. To take account of this, most Universities now offer separate courses in topics like polymer science, ceramic materials, fracture mechanics, composites, superalloys, etc. As is evident, all of these topics are full courses in their own right and to combine all into one subject while still maintaining a University level coverage in terms of depth would be impossible. Given the audience for which this book is intended, emphasis has been placed on metals which the materials, mechanical and production engineers are likely to deal with most of their working lives. It is envisaged that students wishing to study the other topics in greater depth than covered here will take the individual courses. For the same reason, dislocation theory and related topics have not been considered in great detail here.

Throughout this book, illustrations have been used liberally to help amplify the theory. Where applicable, worked examples of numerical problems have been provided. At the end of each chapter, un-worked problems are given (with answers to numerical problems). These questions are adopted mainly from past examination papers taken at Nairobi and Moi Universities. Two or three books recommended for further reading on each topic are given at the end of each chapter.

I would like to thank my colleagues and numerous past students who have contributed in one way or another to the preparation of this book. I wish to mention particularly Dr. Stanley Shitote and Mrs. Joyce Nderitu. Finally, I am grateful to Dr. George Katana of Kenyatta University and Dr. John Kihiu of Jomo Kenyatta University of Agriculture and Technology for their useful comments on the manuscript.

George O. Rading

# LIST OF SYMBOLS

| | |
|---|---|
| a | Atomic spacing |
| | Crack length (surface crack) |
| | Half crack length (centre crack) |
| | Lattice parameter, x-direction |
| apf | Atomic Packing Factor |
| A | Area |
| | Atomic number |
| b | Lattice parameter, y-direction |
| **b** | Burger's vector |
| BCC | Body Centred Cubic crystal structure |
| c | Lattice parameter, z-direction |
| CVN | Charpy V Notch |
| d | Diameter |
| | Average grain diameter |
| $d_{hkl}$ | Interplanar spacing for Miller Indices, (hkl) |
| E | Young's modulus (Modulus of elasticity) |
| $E_c$ | Creep modulus |
| $E_r$ | Relaxation modulus |
| F | Force (interatomic) |
| FCC | Face Centred Cubic crystal structure |
| G | Shear modulus |
| hkl | Miller indices (crystallographic planes) |
| HB | Brinell hardness |
| HCP | Hexagonal Close Packed crystal structure |
| HRx | Rockwell hardness. The x specifies the scale |
| HV | Vicker's hardness |
| k | Boltzmann's constant |
| K | Stress intensity factor |
| $K_c$ | Critical value of K |
| $K_{Ic}$ | Plane strain fracture toughness |
| l | Length |
| $l_c$ | Critical length (fibers) |
| $M_n$ | Molecular weight (number average) |
| $M_w$ | Molecular weight (weight average) |
| n | Strain hardening exponent |
| N | Number of fatigue cycles |
| $N_A$ | Avogadro's number |
| $N_f$ | Fatigue cycles to failure |
| P | Force (mechanical) |

| | |
|---|---|
| Q | Activation energy |
| R | Atomic radius |
| | Stress ratio: minimum stress/maximum stress |
| t | Time |
| $t_r$ | Time to rupture |
| T | Temperature |
| $T_g$ | Glass transition temperature |
| $T_m$ | Melting temperature |
| uvw | Miller indices (crystallographic directions) |
| U | Energy |
| $U_R$ | Modulus of resilience |
| UTS | Ultimate tensile strength |
| V | Potential difference |
| x,y,z | Space coordinates |
| $\alpha, \beta, \gamma$ | Lattice parameters (angles between axes in unit cell) |
| | Phase designations (solid solutions) |
| $\gamma$ | Shear strain |
| $\delta$ | Percent elongation at fracture |
| $\psi$ | Percent reduction in area |
| $\varepsilon$ | Engineering strain |
| $\varepsilon_n$ | Natural (logarithmic or true) strain |
| $\lambda$ | Wavelength of radiation |
| $\eta$ | Viscosity |
| $\nu$ | Poisson's ratio |
| $\theta$ | Diffraction angle |
| $\rho$ | Density |
| $\sigma$ | Direct (normal) stress |
| $\sigma_m$ | Mean stress |
| $\sigma_{max}$ | Maximum stress |
| $\sigma_{min}$ | Minimum stress |
| $\sigma_t$ | True stress |
| $\sigma_Y$ | Yield stress |
| $\sigma_R$ | Stress range: Maximum stress - minimum stress |
| $\tau$ | Shear stress |
| $\sigma_c$ | Critical value of $\sigma$ |

# ABBREVIATIONS OF UNITS

| | |
|---|---|
| A | Ampere |
| C | Coulomb |
| $^o$C | Degrees Celcious |

| eV | Electron volt |
| g | Grams |
| GPa | Giga Pascal |
| J | Joule |
| K | Degree Kelvin |
| kg | Kilogram |
| m | Meter |
| mm | Millimeter |
| mol | Mole |
| MPa | Mega Pascal |
| N | Newton |
| nm | Nanometer |
| P | Poise |
| Pa | Pascal |
| s | Second |
| μm | Micrometer |
| W | Watt |

# CHAPTER ONE

# ATOMIC STRUCTURE AND BONDING

## 1.1  INTRODUCTION
### 1.1.1   Role of Materials Science and Engineering

The field of study known as materials science concerns itself with the generation and application of knowledge that relates the structure and composition of a material to its properties (i.e., structure-property relationship). The properties of the material in turn dictate the uses to which the material may be put. Materials engineering on the other hand seeks to find ways of altering the structure of naturally existing materials so as to achieve a desired set of properties. As will be discussed in subsequent pages, this is achieved through mixing of different materials in specific ways, heat treatment, and by changing the method by which the material is processed into its final shape.

Materials play a very important part in our daily lives. Indeed, technological advancement is so strongly linked to advancement in materials that historical ages are named after materials. Thus, one hears of the Stone Age, the Bronze Age, the Iron Age, etc. Engineers (of whatever specialization) in particular must use materials in their designs and hence must have a thorough comprehension of materials science and engineering to enable then to make a judicious selection of materials for a particular use. This judicious selection requires that engineers take into consideration the service conditions, the economics of the design (i.e., the cost of the material), long term effects (material deterioration), processability, etc. before choosing a material for a given application.

As already stated, the said properties depend on the structure of the material. Structure, here may be defined as the internal arrangement of the components that make the material, and may be viewed at different levels:

(i) The subatomic level: Here, attention is focused on the electrons and their interaction with the nucleus. This level is generally the preserve of physicists and chemists.

(ii) The atomic level: The organization and arrangement of atoms relative to each other.

(iii) Microscopic level: The structure as seen under a light microscope. Here, the phases present and their arrangement as grains (i.e., the grain structure) are studied.

(iv) Macroscopic level: The structure as seen by the unaided (naked) eye.

Materials science and engineering are mainly concerned with the structure at the atomic and microscopic levels.

The term property as used here refers to the response of a material to external stimuli that may be mechanical, electrical, magnetic, thermal, optical or deteriorative. Since this book targets mechanical and production engineers, the major concern will be mechanical properties i.e., the response of a material to external forces applied to it.

## 1.1.2 Classification of Materials

Materials may be classified into three primary classes: metals, polymers and ceramics. The basis of this classification is purely the chemical makeup of the material. Metals are a combination of metallic elements where the atoms are held together by the metallic bond. Ceramics, on the other hand, are compounds of metallic and non-metallic elements in which the bonding is principally ionic (metallic oxides, nitrides, sulphides, carbides, etc.). Polymers are compounds of carbon and other non-metallic elements (hydrogen, nitrogen, oxygen, chlorine, fluorine, etc.) where the bonding is principally covalent.

In addition to the primary materials classes, there are two important secondary (or engineered) classes of materials. These are composites and semiconductors. Composites result when two materials in the primary classes are mixed physically into a single material. Semi conductors have properties in between those of conductors of electricity and those of insulators, and are made by adding minute quantities of selected impurities to certain elements. This class of materials will not be studied further in this book as it is of more interest to electrical engineers.

It is obvious from the above that a study of materials science and engineering must start with a study of the structure. It is further assumed that the reader is familiar with the atomic structure of materials as presented in more elementary courses. For completeness, a brief revision of the major aspects is presented hereafter.

## 1.2  ATOMIC STRUCTURE OF MATTER

All matter is composed of minute fundamental particles called atoms. In solid materials, the arrangement of these atoms and their interactions dictate the properties of the said materials. Each of the atoms is in turn composed of a nucleus surrounded by moving electrons. The nucleus is itself composed of protons and neutrons. Protons and electrons are electrically charged with a charge of 1.60 x $10^{-19}$ C. The neutron is electrically neutral. The mass of the proton is about the same as that of the neutron and is equal to 1.67 x $10^{-27}$ kg. That of the electron is 9.11 x $10^{-31}$ kg.

The number of protons in their nuclei characterizes elements. This number is termed the atomic number, Z, and is equal to the number of protons in the atom. Since Z is a whole number, it varies in integral units from one for hydrogen to 94 for plutonium. (Note: elements with higher atomic numbers have been artificially produced with Z up to 103.)

The mass of an atom is the sum of the masses of the protons and neutrons (the contribution of the electrons to the mass is negligible). This sum is termed the atomic mass number. Atoms of the same element may have different numbers of neutrons. Thus, two atoms of the same element may have different mass numbers while retaining the same atomic number. The phenomenon is termed isotropy and the corresponding atoms termed isotopes. Isotopes exist in specific proportions in naturally occurring elements and a weighted average atomic weight can be calculated. Due to this averaging, the atomic weight may be a non-integral number.

In calculations involving atomic weights, it is convenient to define the atomic mass unit (amu). One amu is $1/12^{th}$ of the atomic mass of carbon 12, which is the most common isotope of carbon. The atomic weight of an element may then be specified in terms of the number of atomic mass units in its atom. Alternatively, the mass per mole of material may be used. A mole of a substance contains 6.023 x $10^{23}$ atoms or molecules. This number is termed Avogadro's number, $N_A$. One atomic mass unit is equivalent to one gram per mole. As such, if the atomic weight of a substance is x, then one mole of the substance weighs x grams. Alternatively, the mass of one amu is $1/N_A$ grams i.e., 1.66 x $10^{-24}$ grams.

## 1.3  ATOMIC MODELS
### 1.3.1  Rutherford-Bohr Atom Model
The behavior of electrons in atoms is best treated by use of principles of quantum mechanics. One consequence of quantum mechanics was the proposal of an atomic model known as the Rutherford-Bohr model. This model proposes that electrons revolve around the nucleus in specific orbitals (figure 1.1). The model stipulates further that energies of electrons are quantinised i.e., that electrons may have only certain specific energies.

When an electron changes its energy, it makes a quantum jump from one orbital (or energy level) to the next.

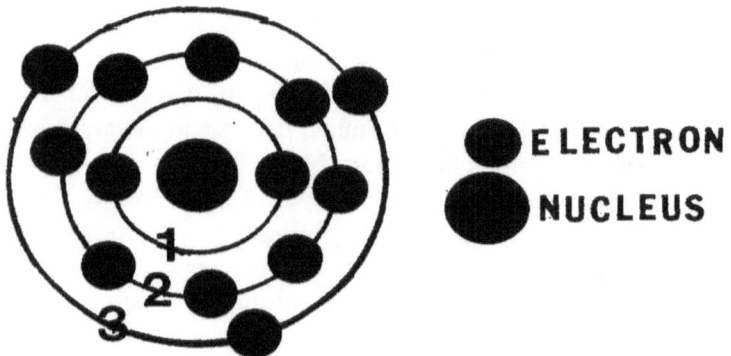

**Fig. 1.1** The Rutherford-Bohr atom model of aluminium

## 1.3.2 Wave Mechanical Model

This model was developed when it was realised that there were some questions that could not be answered by the Bohr atom model. The model considers an electron as having both wave and particle characteristics. The position of the electron around the nucleus is described by a probability function or distribution. Each electron is characterized by a set of four parameters termed quantum numbers. Three of these numbers specify the size, shape and orientation of an electron's probability distribution density. The first quantum number, (the principal quantum number, n) specifies the shell in which the electron is situated. It (i.e., n) may be given as an integer: 1, 2, 3, etc., or a letter: K, L, M, N, etc. Thus, if an electron is in shell K its principal quantum number is 1. It should be noted that the first shell corresponds to the first orbital in the Rutherford-Bohr model, and n = 2 corresponds to the second orbital, etc.

Within each shell, there exists various sub shells. The second quantum number, l, specifies this sub shell. Lower case letters are used to specify the sub shells as follows: s (l = 0); p (l = 1); d (l = 2); f (l = 3). The permissible number of sub shells depends on the principal quantum number as shown in table 1.1.

Within each of the sub shells, there are different energy states. This is specified by the third quantum number, $m_l$ that takes values of 0; +1; -1; etc. Finally, each electron is associated with a spin moment. The spin moment, $m_s$, is specified by the fourth quantum number, which may have the values $+^1/_2$ or $-^1/_2$ depending on the spin orientation.

**Table 1.1** Electron states in various electron shells.

| | sub shell | # of electron states | # of electrons per sub shell | # of electrons per shell |
|---|---|---|---|---|
| K | s | 1 | 2 | 2 |
| L | s | 1 | 2 | |
| | p | 3 | 6 | 8 |
| M | s | 1 | 2 | |
| | p | 3 | 6 | |
| | d | 5 | 10 | 18 |
| N | s | 1 | 2 | |
| | p | 3 | 6 | |
| | d | 5 | 10 | |
| | f | 7 | 14 | 32 |

The lower the principal quantum number, the lower the energy level associated with an electron. Thus the energy level 1s is lower than the energy level 2s. Similarly, within each shell, the energy level increases with the second quantum number, l. Therefore, the energy of 4p is lower than that of 4f. However, there may be an overlap in the energy levels of adjacent shells. As a consequence, the energy level of 4s is lower than that of 3d. Note that the lowest energy levels are filled with electrons before those with higher energy states. Furthermore, filling of energy levels is controlled by Paulli's exclusion principle, which states that each energy state can be filled with no more than two electrons, which must have different spins. In other words, no two electrons can have the same set of four quantum numbers.

An atom is said to be in its ground state if electrons fill the lowest possible energy levels. The manner in which the energy levels are occupied describes the electron configuration or structure of the atom. In specifying the electron structure, the number of electrons occupying each sub shell is given as a superscript. As an example, the electron structure of carbon, with an atomic number of six is: $1s^2 2s^2 2p^2$.

The electrons that occupy the outermost shell are termed the valence electrons. These are the electrons that participate in atomic bonding and hence, dictate the physical and chemical properties of the element concerned if it is a solid. Atoms in which the outermost shell is completely filled are said to have a stable electron configuration. Such atoms have no tendency to take part in a chemical reaction and are termed inert gases (helium, neon, argon, etc.). For this reason, the stable electron configuration is also termed the inert gas structure. Atoms that do not have the stable electron configuration attempt to attain the same by sharing or exchanging electrons. This is the reason for chemical reactions between elements.

## 1.4  THE PERIODIC TABLE

A classification of all elements according to their electron configuration is termed the periodic table (see Appendix). The elements are arranged in order of increasing atomic number in horizontal rows or periods. All elements in the same column or group have the same number of valence electrons. As a result, they have similar chemical properties. Elements in group IA for example have one valence electron, etc. Three of the periods are longer than the rest. The extra groups are designated IIIB, etc. These elements are termed transition elements. They have partially filled 3d electron states.

Elements that are to the left of the periodic table may give up their valence electrons easily to form positively charged ions. They are said to be electropositive elements or metals. Those to the right of the periodic table can acquire electrons to form negatively charged ions. These are said to be electronegative.

Example 1.1: Using Table 1.1, state the electron configuration of chlorine, that has an atomic number of 17.

Answer: $1s^2 2s^2 2p^6 3s^2 3p^5$

Example 1.2: To which group of the periodic table do the elements with the following electron configurations belong? State also the atomic number of the element concerned.

(a) $1s^2 2s^2 2p^1$
(b) $1s^2 2s^2 2p^6$
(c) $1s^2 2s^2 2p^6 3s^2$
(d) $1s^2 2s^2 2p^6 3s^2 3p^6 3d^1 4s^2$

(e) $1s^2 2s^2 2p^6 3s^2 3p^6 3d^{10} 4s^2 4p^5$

Answers:

(a) The outermost shell (shell L) has 3 electrons => the group is IIIA (Note: the answer cannot be IIIB since the M shell has not started filling).

The atomic number is simply the sum of the number of electrons i.e., $2 + 2 + 1 = 5$.

(b) The outermost shell (shell L) has 8 electrons => this must be an inert gas in group O.

$Z = 2 + 2 + 6 = 10$.

(c) The outermost shell is shell M. It has 2 electrons => group is IIA.

$Z = 2 + 2 + 6 + 2 = 12$.

(d) It is noted that the outermost shell is N ($n = 4$) and has two electrons. However, the 3d sub shell is not filled since it needs 10 electrons to fill it. => This is a transition metal => group is IIIB (Note: the single 3d electron is also a valence electron).

$Z = 2 + 2 + 6 + 2 + 6 + 1 + 2 = 21$.

(e) Here, shell N has 7 electrons while shell M is completely filled. The group is VIIA.

$Z = 35$.

Example 1.3: Quantum numbers for each electron are given as: n l $m_l$ ($m_s$) e.g., 100(- $1/2$). Give the quantum numbers for the valence electrons of magnesium.

Answer: The electron configuration of magnesium is:

$1s^2 2s^2 2p^6 3s^2$

i.e., the valence electrons are in the third shell => $n = 3$; the sub shell is s (i.e., $l = 0$); s has only one energy state hence $m_l = 0$; and $m_s = +1/2$ or $-1/2$. Thus the quantum numbers for the two electrons are:

$300 (+ 1/2)$ and
$300 (- 1/2)$

## 1.5 ATOMIC BONDING

Atomic bonding, which is responsible for formation of solids, is due solely to attractions between atoms. There are two main types of bonds:

(i) Primary bonds: These are strong bonds that result from exchange or sharing of electrons, and may be covalent, ionic or metallic bonds.

(ii) Secondary bonds: These are relatively weak bonds holding molecules (i.e., they are intermolecular) together. In the case of inert gases in liquid or solid state, these bonds also hold the atoms together.

### 1.5.1 Ionic Bonds

Atoms become very stable when their outer electron shells are completely filled. Atoms to the left of the periodic table (e.g., sodium, Na) can achieve this by losing the outermost electrons, while those to the right (e.g., chlorine, Cl) achieve the same by gaining one or two electrons. However, the atoms left behind gain net positive ($Na^+$) or negative ($Cl^-$) charge and are then called ions. The attraction between the ions constitute the interatomic bond in the compound e.g., sodium chloride, NaCl. As the ions approach each other however, there is repulsion between their electron clouds. The optimum distance apart, $a_o$ (see figure 1.2) corresponds to the point where the resultant force is zero (or resultant energy, E, is minimum) and is the value of a to make the expression in equation 1.1 a minimum.

$$E = \frac{(qz_1)(-qz_2)}{a} + \frac{q^2 b}{a^n} \qquad (1.1)$$

In equation 1.1, q = electron charge; $z_1$ and $z_2$ = the number of ions of each type making the compound; b and n (n = 5 to 12) are constants.

Other pairs of atoms lose more than one electron depending on the valence. This type of bond is called ionic bond and results from exchange of electrons.

Ionic compounds dissociate in solution and the resultant mobility of the ions accounts for the ability of such solutions to conduct electricity.

Example 1.4: The potential energy between two ions forming an ionic bond may be represented by:

$$E = -\frac{A}{a} + \frac{B}{a^n}$$

where A, B and n are constants, while a is the distance between the centers of the ions. Derive expressions for the equilibrium potential energy and the equilibrium distance apart, $a_0$.

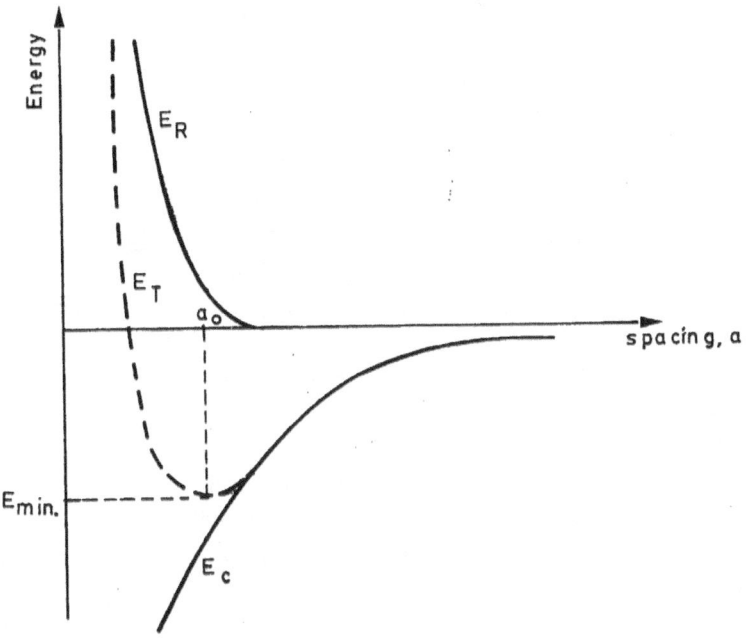

**Fig. 1.2** The potential energy of two approaching ions with unlike charges. $E_R$ = electron repulsion energy, $E_c$ = energy due to Coulombic attraction, $E_T$ = total energy.

Answer:
Differentiate the above equation with respect to a and set the result to zero:

$$\frac{dE}{da} = \frac{A}{a^2} - \frac{nB}{a^{n+1}} = 0$$

$$\Rightarrow \frac{A}{a_o^2} = \frac{nB}{a_o^{n+1}}$$

9

$$a_o = \left[\frac{A}{nB}\right]^{1/(1-n)}$$

The minimum energy corresponds to $a_o$. Substituting $a = a_o$ in the original equation:

$$E_o = -\frac{A}{a_o} + \frac{B}{a_o^{\,n}}$$

$$E_o = -\frac{A}{(\frac{A}{nB})^{1/(1-n)}} + \frac{B}{(\frac{A}{nB})^{n/(1-n)}}$$

## 1.5.2 Covalent Bonds

This type of bond involves the sharing of electrons to achieve the inert gas structure. As two atoms approach each other, the valence electrons of one are attracted to the nucleus of the other (but repelled by electrons of the other) and there is an overlap of the two electron clouds. This overlap constitutes the bond. It is responsible for holding atoms together in gaseous molecules e.g., $O_2$, $N_2$, $Cl_2$. In solids it is mainly found in elements at the centre of the periodic table e.g., C, Si, Ge.

Since specific atoms share electron clouds, the covalent bond is stereospecific (except in the benzene ring). The bonds are also directional due to repulsion between the electron clouds (hence their tendency to stay as far apart as possible). Covalent bonds are very strong as exemplified by diamond.

## 1.5.3 Metallic Bonds

Elements to the left of the periodic table, e.g., lithium, sodium, aluminium, etc. (which are termed metals) have their valence electrons shielded from attraction of the nucleus by the inner electron shells. When many atoms of these elements come together, these electrons are lost and form an "electron cloud" which is shared by all the atoms. The shared "cloud" constitutes the bond.

This type of bonding accounts for several characteristic properties of metals:

1. Electrical conductivity: Since the electrons are easily moved an applied potential difference easily move them, constituting electric current. This makes metals good conductors of electricity and heat.

2. Lustre and opacity: The electron cloud interacts with photons of light accounting for the good lustre of metals. The same interaction accounts for the opacity of metals since no light can pass through.

3. Ductility/malleability: Due to the shared electron cloud, neighbouring atoms can easily change neighbours without any bonds being broken. For this reason, metals have good ductility and are malleable i.e., they deform easily without breaking.

## 1.5.4 Secondary Bonds

These arise from internal dipoles. The bond results from coulombic attractions between the dipoles. There are two types:

Permanent dipoles: When a covalent bond (in a molecule) is formed between atoms of differing electronegetivity, the shared electrons move closer to one atom than the other. This renders that end of the molecule slightly negative. The other end becomes slightly positive and the molecule becomes a dipole. The attraction between the resulting dipoles constitutes the bond.

When hydrogen is one of the atoms forming the molecule, this type of bond is fairly strong and is called hydrogen bond or bridge. The reason for this is that when hydrogen's lone electron is shared, one end of the resulting molecule is a bare proton and hence has a fairly strong positive charge. Such compounds e.g., $H_2O$, HCl, are usually liquid at room temperature. (See figure 1.3.)

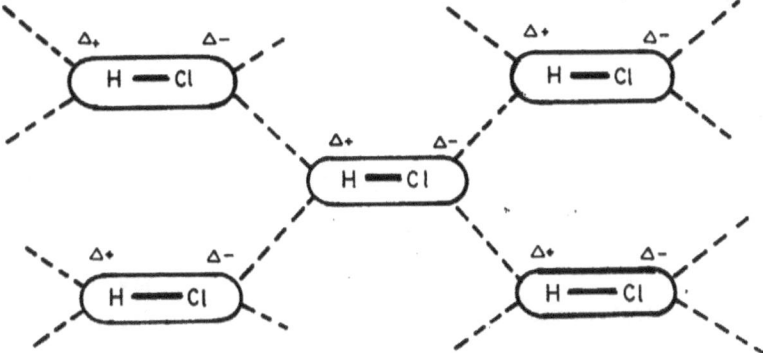

**Fig. 1.3** Schematic representation of the hydrogen bond.

Temporary dipoles: Statistical irregularities in electron distribution occur in molecules and may render one end of a molecule (between atoms of the same electronegetivity) temporarily negative. The resulting dipole will induce dipoles in neighbouring molecules and the dipole-dipole attraction constitutes the bond. (See figure 1.4). This is the type of bond that holds the molecules together in liquefied (or solidified) gases. The dipole-dipole attraction can, of course, only overcome kinetic energy forces at very low temperatures. Moreover, there is fluctuation in the sign of the charge at any end of the dipole and hence these dipoles are termed as fluctuating dipoles.

Temporary dipoles are responsible for holding inert gases together in the liquid state. When they occur in inert gases, temporary dipoles are also called Van der Waals bonds.

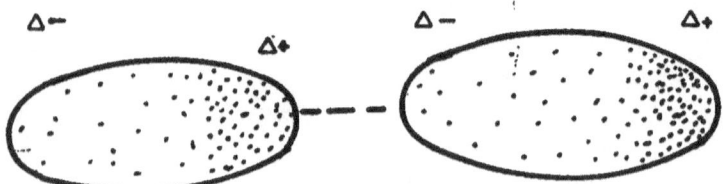

**Fig. 1.4** Schematic representation of a fluctuating dipole

## PROBLEMS

1.1 Which properties of a metal depend mainly on the electron configuration of its atoms?

1.2 Explain clearly what is meant by hydrogen bond and Van der Waals bond.

1.3 What type of bond would you expect to form between two atoms of an element having the following atomic numbers: 8; 10; 12? (Ans: Covalent; Van der Waals; metallic).

1.4 Explain, in terms of the type of bonding, why metals have good lustre and good electrical conductivity.

1.5 What determines the distance between ion centres in an ionic compound? Illustrate your answer with a sketch.

1.6 Refer to example 1.4. Given that for a NaCl compound, A = 1.436, B = 7.32 x $10^{-6}$, and n = 8 when E is in electron volts and a is given in nanometers, determine, both analytically and graphically, the equilibrium spacing and the equilibrium potential energy. Calculate also the force of attraction between the ions at the equilibrium spacing. (Ans: 0.236 nm, -5.3 eV, 4.1 x $10^{-9}$ N.)

1.7 With the aid of sketches, explain the mechanisms of two types of intermolecular bonds.

## Further Reading

Anderson, J. C. Leaver, K.D. Rawlings, R.D. and Alexander, J.M. <u>Materials Science,</u> 3rd. ed. ELBS and Van Nostrand Reinhold, Berkshire, 1985.

Callister, W.D. <u>Materials Science and Engineering,</u> 3rd ed. John Wiley and Sons, New York, 1994.

Pascoe, K. J. <u>An Introduction to the Properties of Engineering Materials</u>, 3rd. ed. ELBS and Van Nostrand Reinhold, Berkshire, 1982.

# CHAPTER TWO

# THE CRYSTAL STRUCTURE OF MATERIALS

## 2.1  CRYSTAL PATTERNS

As atoms or molecules are coordinated and bonded with each other to form solids specific patterns may emerge. If a solid exhibits long-range order, it is called a crystal. There is a shape, the crystal lattice, which characterizes the order in a crystal e.g., in a 2-dimensional case; the lattice may be square, rectangular, etc. In 3-dimensions, there are 3 sides and 3 angles between the sides. Thus, the seven possible space lattices can be constituted by variation of the lengths of the sides (a, b and c) and the angles between different planes ($\alpha$, $\beta$, $\gamma$) as follows:

| | | |
|---|---|---|
| Cubic | $a = b = c;$ | $\alpha = \beta = \gamma = 90°$ |
| Tetragonal | $a = b \neq c;$ | $\alpha = \beta = \gamma = 90°$ |
| Orthorhombic | $a \neq b \neq c;$ | $\alpha = \beta = \gamma = 90°$ |
| Monoclinic | $a \neq b \neq c;$ | $\alpha = \gamma = 90°; \beta \neq 90°$ |
| Triclinic | $a \neq b \neq c;$ | $\alpha \neq \beta \neq \gamma \neq 90°$ |
| Hexagonal | $a = b \neq c;$ | $\alpha = \beta = 90°; \gamma = 120°$ |
| Rhombohedral | $a = b = c;$ | $\alpha = \beta = \gamma \neq 90°$ |

The space lattice is usually defined by a set of equivalent points (say at the corners of each of the above shapes) called lattice points. Crystal structures are obtained by putting atoms (or compounds) at the lattice points. It is possible to identify the smallest unit, which has the characteristics of the whole lattice (i.e., the lattice may be considered to have been built up by stacking these units) called the unit cell. The dimensions of the unit cell, a, b, and c are called the lattice constants of the crystal.

There are variations possible within the crystal systems and hence 14 space lattices are possible as shown in figure 2.1. The 14 space lattices are called Bravais lattices.

## 2.2  METAL CRYSTALS

Because they have given up their valence electrons in the process of bond formation, metal atoms (or ions) are approximately spherical. Therefore,

metal crystals are like spheres of equal size stacked together. This favours
the formation of a cubic lattice and hence most metal crystals are cubic. The
common metal crystals are:

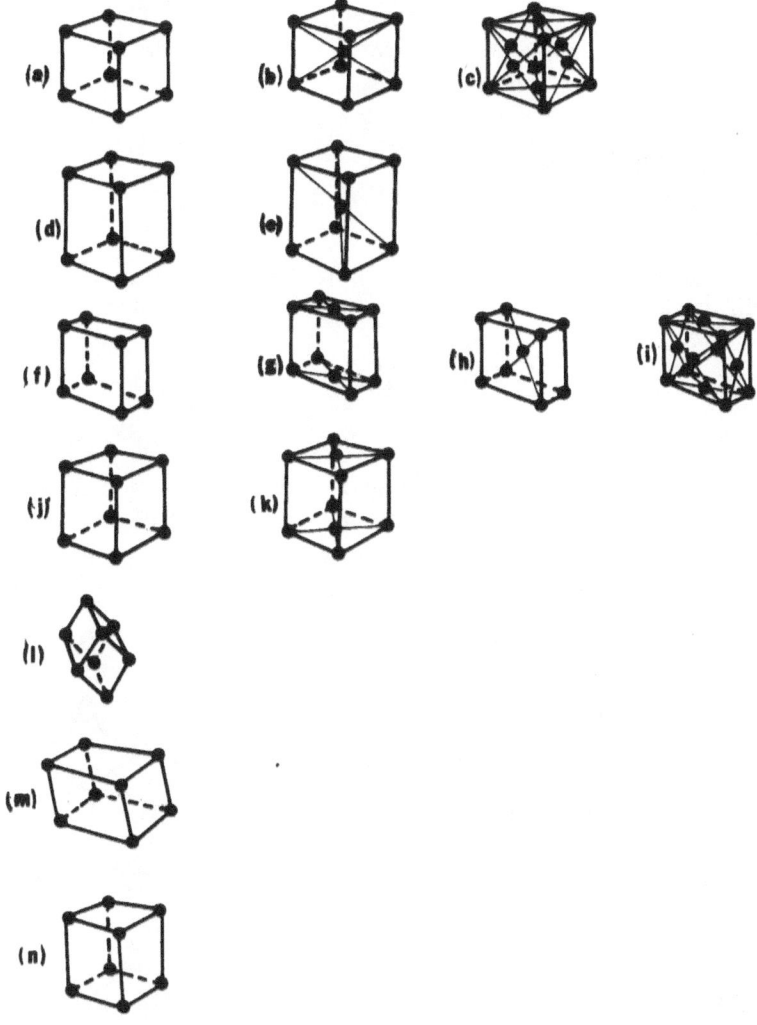

**Fig. 2.1** The 14 Bravais lattices: (a) simple cubic, (b) body centred cubic, (c)
face centred cubic, (d) tetragonal, (e) body centred tetragonal, (f)
orthorhombic, (g) end centred orthorhombic, (h) body centred
orthorhombic, (i) face centred orthorhombic, (j) monoclinic (k) end centred
monoclinic, (l) triclinic, (m) rhombohedral (n) hexagonal.

## 2.2.1 Simple Cubic (SC)

Atom centres are placed at the 8 corners of the cube as shown in figure 2.2. The coordination number (number of nearest neighbours) is 6, and there is only 1 atom per unit cell (each atom is shared by 8 neighbouring unit cells hence there are $8 * 1/8 = 1$ atom per unit cell).

The closeness of the packing is quantified by the atomic packing factor, apf, defined as:

$$apf = \frac{volume\_of\_atoms}{volume\_of\_unit\_cell} \qquad (2.1)$$

If R = radius of an atom, then for simple cubic structure, a = 2R (see figure 2.2 (b)) and:

$$apf = \frac{4/3\pi R^3}{a^3} = \frac{4/3\pi R^3}{(2R)^3} = 0.52 \qquad (2.2)$$

Due to the low packing (more energy is lost when the packing is closer making closer packed structures more stable) no common metal crystallizes with a simple cubic crystal structure.

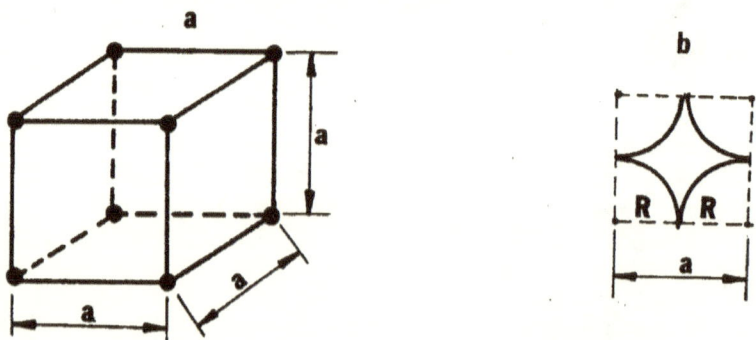

**Fig. 2.2** The simple cubic crystal structure: (a) unit cell; (b) geometry of a cube face.

## 2.2.2 Body Centred Cubic (BCC)

The structure is similar to the simple cubic but with an extra atom at the centre of the cube (figure 2.3 (a)). The coordination number is 8, as can be ascertained by considering the atom at the cube centre. There are 2 atoms/unit cell ($1/8 \times 8 = 1$ as in the case of the SC structure plus the atom at the cube centre which is not shared by any other unit cell). From figure 2.3 (b), $a = 4R/ 3^{1/2}$. Hence:

$$apf = \frac{2 \times 4/3\pi R^3}{(4\dfrac{R}{\sqrt{3}})^3} = 0.68 \qquad\qquad (2.3)$$

The common metals with this structure are chromium, α-iron, δ-iron, vanadium, molybdenum, tungsten and β-titanium.

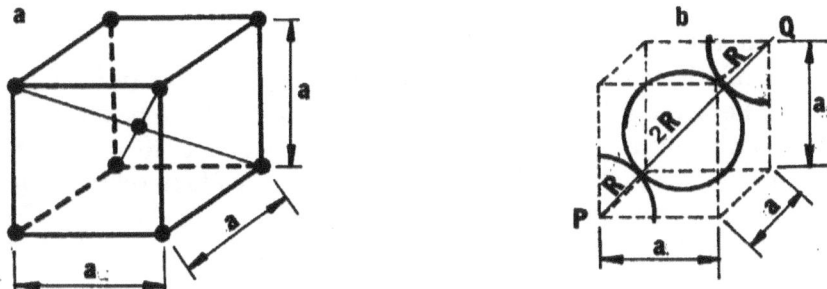

**Fig. 2.3** Body centred cubic crystal structure: (a) unit cell (b) relation between a and R.

### 2.2.3 Face Centred Cubic (FCC).

This structure is based on the simple cubic structure but with additional atom centres placed at each of the six faces of the cube as shown in figure 2.4 (a). The coordination number is 12 that is the highest possible and there are 4 atoms/unit cell (each of the six atoms on the face are shared by 2 unit cells => atoms/uc = $1/8 \times 8 + 1/2 \times 6 = 4$). From figure 2.4 (b), $a = 4R/2^{1/2}$ and:

$$apf = \frac{4 \times 4/3\pi R^3}{(4R/\sqrt{2})^3} = 0.74 \qquad\qquad (2.4)$$

The common metals with this crystal structure include γ-Fe, Ni, Cu, Al, Ag, Au, and Pb.

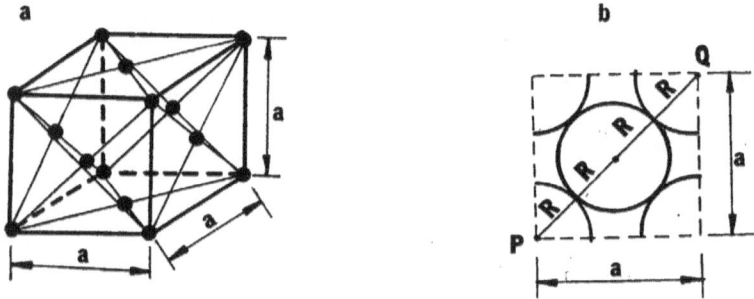

**Fig. 2.4** Face centred cubic crystal structure (a) unit cell (b) relation between a and R.

## 2.2.4 Hexagonal Close Packed (HCP)

The structure is shown in figure 2.5 (a). The CN is 12 (consider atom at centre of hexagon. It has 6 neighbours in the same plane, 3 below and 3 above). Two lattice constants, (a) and (c) are required to define the dimensions of the unit cell. There are 6 atoms/unit cell [3 (u cell centre) + $1/2$ x 2 (hexagon centre) + $1/6$ x 12 (hexagon corner) = 6].

The apf can be calculated by reference to figure 2.5 (b):

$$AD = \sqrt{(4 R^2 - R^2 )} = R\sqrt{3} \text{ (2.5)}$$

Area of hexagon base = 6 x Area ABC = 6 x $1/2$ x Rx3$^{1/2}$ x 2R = 6 x 3$^{1/2}$ x R$^2$

$$OA = \frac{2}{3} AD = \frac{2}{\sqrt{3}} R$$

(Since atom F lies at the centroid of ABC)

$$(OF)^2 = (AF)^2 - (OA)^2 = (2R)^2 - (\frac{2}{\sqrt{3}} R)^2$$

$$OF = \sqrt{8/3}R = \frac{2\sqrt{2}}{\sqrt{3}} R$$

$$c = 2(OF) = \frac{4R\sqrt{2}}{\sqrt{3}} \qquad\qquad (2.6)$$

Vol. of unit cell = area of base x c. And:

$$apf = \frac{6x4/3\pi\ R^3}{6\sqrt{3}\ R^2\ x4R\sqrt{2/3}} = 0.74 \qquad (2.7)$$

Among the common metals with HCP structure are magnesium, Zinc, Cobalt, Beryllium, zirconium, and α-titanium.

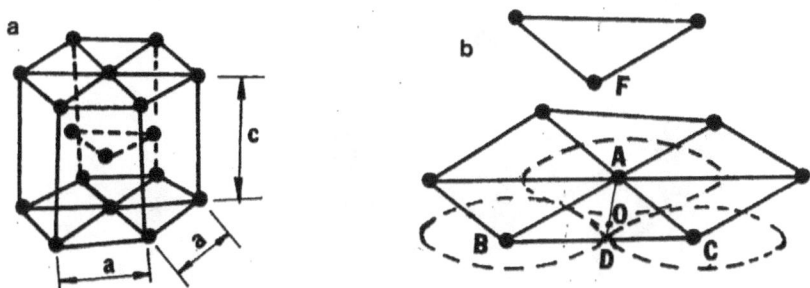

**Fig. 2.5** HCP Crystal Structure (a) unit cell (b) crystal geometry

Example 2.1: Sketch the unit cells of molybdenum and nickel and calculate the atomic packing factor of each.

From above, molybdenum has a BCC crystal structure. The unit cell is as shown in figure 2.3 (a). The apf is 0.68 as already calculated. Similarly, nickel has an FCC crystal structure. The unit cell is shown in figure 2.4 (a) and the apf is 0.74.

Example 2.2: Figure E2.2 (a) shows the unit cell of some hypothetical metal. (a) What is the crystal system of the metal and what would the crystal structure be called? (b) If the atomic weight of the metal is 114 amu, what is the theoretical density of the metal in kg/m³? How would the value of density calculated compare with the actual density? (c) Calculate the apf of the unit cell.

(a) From the figure, it is noted that a = b, but not = c, and that α = β = γ = 90°. From section 2.1, the crystal system is tetragonal.

There is an atom at the tetragon centre, hence the crystal structure is body centred tetragonal.

(b) A unit cell was described as the smallest unit having all the characteristics of the whole crystal. Hence the density of the metal should equal that of the unit cell. Therefore:

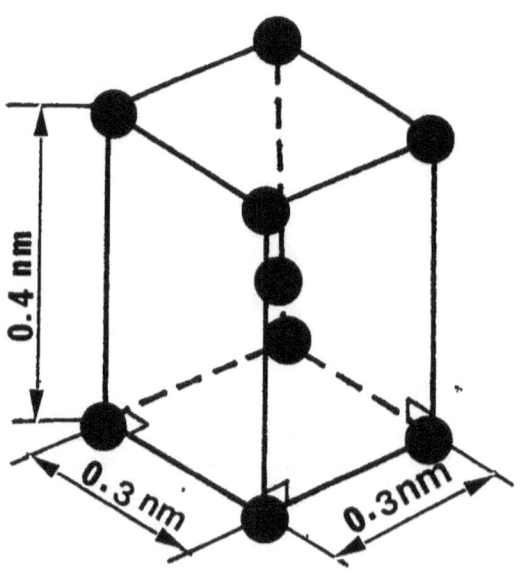

**Fig. E2.2 (a)** A hypothetical unit cell

$$\rho = \frac{mass\_of\_unit\_cell}{volume\_of\_unit\_cell} = \frac{mass\_of\_uc}{0.3x0.3x0.4(nm)^3}$$

There are two atoms per unit cell (the atom at the centre + $1/8$ x 8 atoms at tetragon corners), each weighing 114 amu, or 114 x 1.66 x $10^{-24}$ g. Hence:

$$\rho = \frac{2x114x1.66x\,10^{-24}\,x\,10^{-3}}{36x\,10^{-30}}\,kg/\,m^3$$

$$= 10,515 \text{ kg/m}^3$$

The actual density may differ slightly from the above value due to impurities and imperfections in the crystal.

(c) From the definition of apf given in the text, and noting that there are two atoms per unit cell:

$$apf = \frac{2x4/3\pi\,R^3}{axbxc}$$

To get the value of R, it is observed that the diagonal of the cube is the close packed direction i.e., the atoms touch each other in this direction. From the figure below:

$$PQ = 4R = [0.3^2 + 0.3^2 + 0.4^2]^{1/2} => R = 0.146 \text{ nm}$$

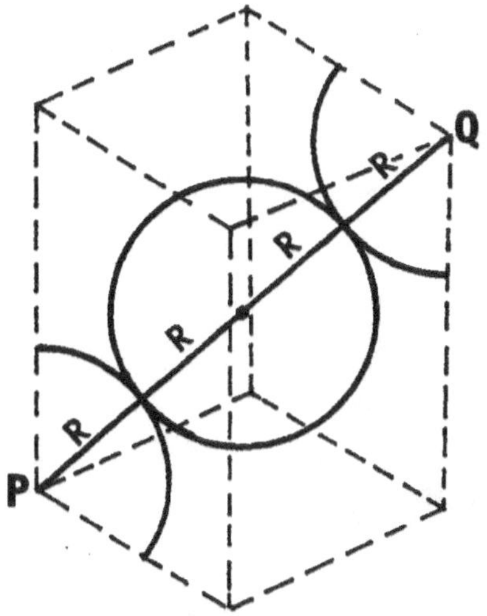

**Fig. E2.2 (b)**

After substituting a = b = 0.3 nm, c = 0.4 nm, and R = 0.146 nm into the equation for apf: apf = 0.724

The crystal structures considered so far may also be visualized as alternate layers of planes of spheres. In the case of BCC and SC crystal structures, the planes have a square arrangement of atoms or spheres (see figure 2.6). To get the SC crystal structure, the next plane is placed such that its atoms are on top of the atoms of the lower plane, and this is repeated throughout. Since each plane is placed similar to all other planes, this may be thought of as an AAAAA stacking of planes. Consider the case in which the second plane is placed such that its atoms are at the centroid of four atoms of the lower plane. The third plane is placed such that the atom centres are at the centroids of the second plane (and hence coincident with the centres of atoms in the first layer), and the pattern is repeated. This arrangement can be thought of as an ABABAB stacking sequence and results in the BCC crystal structure. Those reading this for the first time are encouraged to try this practically for better understanding. Furthermore, this should make it clear that the atom at the centre of the BCC unit cell is no different from the other atoms.

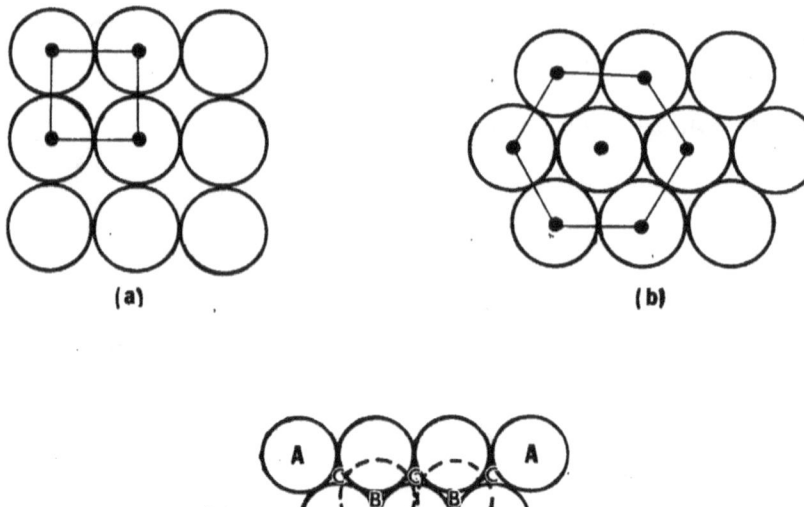

**Fig. 2.6** Stacking of crystal planes (a) A "square" arrangement of atoms within a plane. (b) A "hexagonal" arrangement. (c) Positions of the "B" and "C" centroids in ABCABC stacking.

In the case of FCC and HCP crystal structures, the spheres have a hexagonal arrangement within the plane (figure 2.6). The planes are said to be close packed (see next section). The next plane of atoms is placed such that its atoms are at the centroid of three atoms of the lower plane. Then in the case of HCP, the third layer is placed such that its atoms are directly above those of the first layer (ABABAB stacking). Each hexagon has six "centroids" into which atoms could fit (figure 2.6) but only three can be simultaneously filled. To obtain the FCC crystal structure, the third plane has its atoms corresponding to the three "centroids" which were not filled by the second layer. Hence, it is only the forth plane whose atoms correspond to those of the first plane (called ABCABCABC stacking). The reader should keep in mind that in the case of the FCC crystal structure, the close packed planes are not the cube faces but {1 1 1} planes (see section 2.4 for nomenclature of planes). Once again, practicing with models makes the understanding easier especially for the FCC crystal.

## 2.3 POLYMORPHISM

Some elements exhibit different crystal structures such that each structure exists over well-defined physical conditions. The phenomenon is called polymorphism and each structure is termed a polymorph. (When polymorphism occurs in metals, the word allotropy is also used.) Examples are: carbon which may exist as diamond (diamond cubic crystal structure) or graphite (hexagonal crystal structure); iron, which has the temperature based polymorphs $\alpha$-Fe (BCC), $\gamma$-Fe (FCC) and $\delta$-Fe (BCC); and titanium which may have a BCC crystal structure ($\beta$-Ti) or a HCP crystal structure ($\alpha$–Ti). More about the polymorphism of iron will be said in chapter six and that of titanium in chapter eight.

## 2.4 CRYSTAL PLANES AND DIRECTIONS

From the previous sections, it may be observed that the distance between successive atoms differs in different directions and that some planes are close packed while others are not. The strength of the bond between atoms depends on the distance between atoms and hence crystals are anisotropic (i.e., have different properties in different directions). Polycrystalline materials are however, isotropic (or quasi-isotropic) if the arrangement of their crystals is random and anisotropic if the crystals have a preferred orientation.

It is therefore important to be able to identify crystal planes and directions and the most often used system is the Miller index system. Here, after drawing the unit cell, an origin is chosen at one corner of the unit cell and three perpendicular axes, x, y and z drawn from it. Then, for planes, the intercepts of the plane on the axes u.a, v.b, and w.c (where a, b and c are the lattice constants) are determined. The Miller indices (hkl), of the plane are then $1/_u$, $1/_v$, $1/_w$. For example the plane shown in the last sketch of figure 2.7 intercepts the x axis at $1/_2$a, the y-axis at 1 b and z-axis at infinity. Hence the Miller indices are (2 1 0) (the indices are reduced to the nearest whole number). The indices for individual planes are given in ordinary brackets and with no commas i.e., (h k l), and negative numbers are indicated with a bar over the numeral.

All parallel planes have the same index. Furthermore, some planes, though not parallel, are crystallographically equivalent, that is, the arrangement of atoms on the planes is the same. As an example, consider the six faces of the cube in a cubic system: (1 0 0), ($\bar{1}$ 0 0), (0 1 0), (0 0 1), (0 $\bar{1}$ 0), (0 0 $\bar{1}$). These families of planes are represented in curly brackets as {h k l}. In the above case, the family of planes would be represented as {1 0 0}.

To make calculations involving planes in the hexagonal unit cells easier, four Miller indices, (h k i l), corresponding to the four axes shown in figure 2.5 (a), are specified instead of the three specified for the cubic system.

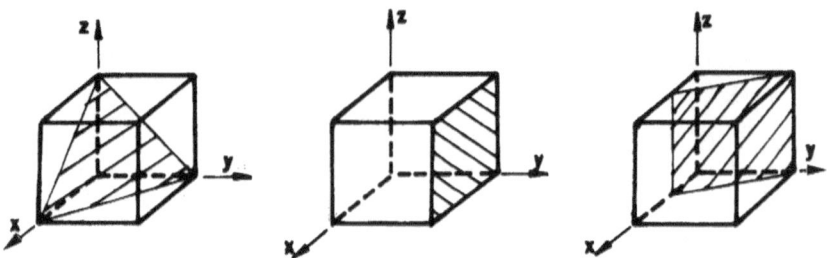

**Fig. 2.7** The (111), (010) and (210) planes in a cubic crystal lattice

Crystallographic directions are similarly indexed: the coordinates of any point through which the direction passes are taken and reduced to the lowest integers. The direction is then given in square brackets as [u v w]. For example, the indices of a direction passing through ($1/4$, $1/2$, 1) are [1 2 4]. Crystallographically equivalent directions are given in pointed brackets: <u v w>.

The Miller index notation makes calculations involving planes and directions easy. From the law of cross product in vector geometry, the line of intersection of two planes ($h_1 k_1 l_1$) and ($h_2 k_2 l_2$) is given by:

$$(h_1k_1l_1) \times (h_2k_2l_2) = [u\ v\ w] \qquad (2.8)$$

i.e., $u = k_1 l_2 - l_1 k_2$; $\qquad v = l_1 h_2 - h_1 l_2$ and, $\qquad w = h_1 k_2 - k_1 h_2$

Planes have the same Miller indices as the directions normal to them and hence if [u v w] lies in plane (h k l) the dot product (h k l).[u v w] = 0, since the plane normal is perpendicular to [u v w] making the cosine of the angle between the two zero.

The angle, $\theta$, between two directions ($u_1\ v_1\ w_1$) and ($u_2\ v_2\ w_2$) can be found from the dot product rule i.e.,

$$\cos\theta = \frac{u_1 u_2 + v_1 v_2 + w_1 w_2}{\sqrt{(u_1^2 + v_1^2 + w_1^2)(u_2^2 + v_2^2 + w_2^2)}} \qquad (2.9)$$

Example 2.3: Show the (2 2 1), (1 0 1) and ($\bar{1}$ 1 0) planes in a cubic lattice.

For a plane with indices (h k l), the intercepts on the axes are $(1/_h.a$ , $1/_k.b, 1/_l.c)$. Moreover, for a cubic crystal lattice, $a = b = c$.

=> For (2 2 1), intercepts are $(1/_2a, 1/_2a, 1a)$
     For (1 0 1), the intercepts are (1a, infinity, 1a)
     For ($\overline{1}$ 1 0), the intercepts are (-1a, 1a, infinity)
The planes are shown in figure E2.3.

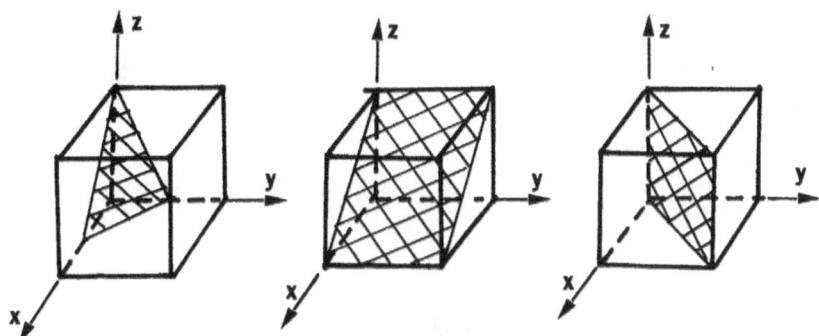

**Fig. E2.3** (2 2 1), (1 0 1) and ($\overline{1}$ 1 0) planes

Note that in the last case, the plane shown is parallel to the one obtained from considering the intercepts given above. This is in order since it was stated above that all parallel planes have the same Miller indices.

The linear density of equivalent sites in a direction is a measure of how close atoms are to each other in the given direction and is given by $1/b_{uvw}$, where $b_{uvw}$ = shortest distance between equivalent sites e.g., atom centres in the direction concerned. The direction with the highest linear density is termed the close packed direction (usually the direction in which the atoms are in contact). It is also possible to get the planer density of atoms in a plane. This may be expressed either as a fraction (the proportion of the area occupied by atoms), or in terms of atoms per unit area (see example 2.4).

Example 2.4: Figure E2.4 (a) shows three different crystallographic planes for a unit cell of some hypothetical metal. (a) Determine the crystal system to which the unit cell belongs. (b) What would the crystal structure be called? (c) Show the [$\overline{1}$ 1 0] direction in the unit cell and calculate the shortest distance between equivalent sites in this direction. (d) Determine the planer density of the (1 0 1) plane as a fraction, and in terms of atoms per square meter.

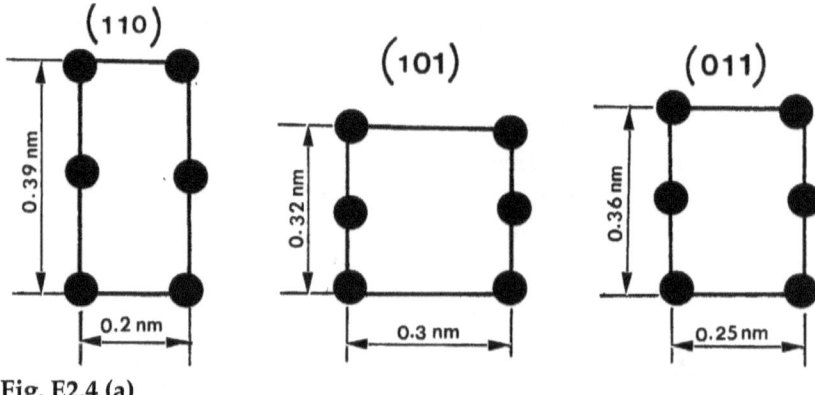

**Fig. E2.4 (a)**

(a) From the planes provided, the unit cell may be constructed as shown in figure E2.4 (b).

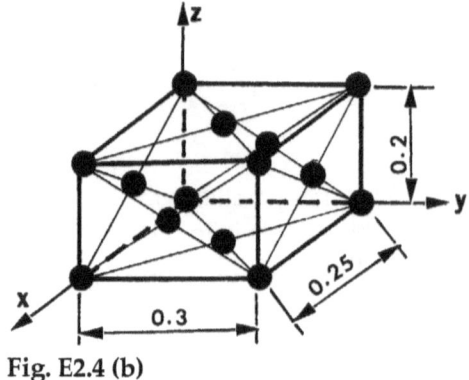

**Fig. E2.4 (b)**

It may be noted that each of the three planes shown gives us at least one lattice parameter. Therefore:

a = 0.25 nm,  b = 0.3 nm,    c = 0.2 nm

Since the lattice constants are all different and all the angles are right angles, the crystal system is orthorhombic.

(b) It is noted that there is an atom at each face centre. Hence the crystal structure is face centred orthorhombic.

(c) From the explanation provided in the text, the [$\bar{1}$ 1 0] direction is a line from the origin passing through point (-1, 1, 0), or any other direction parallel to it. The direction is shown in figure E2.4 (c).

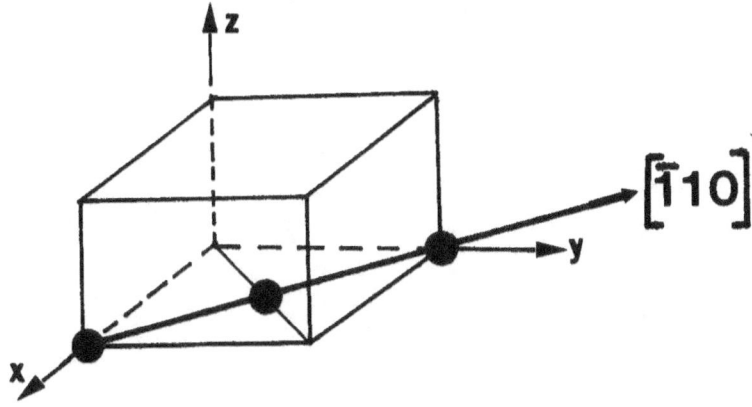

**Fig. E2.4 (c)**

From figure E2.4 (c), the shortest distance between equivalent sites, $b_{[\bar{1}10]}$ is half the diagonal of the bottom face:

$$b_{[\bar{1}10]} = \frac{1}{2}\sqrt{0.3^2 + 0.25^2} = 0.20nm$$

The linear density of equivalent sites in this direction is then:

LDES = $1/b$ = 5 sites per nm = 5 x $10^9$ sites per meter

(d) To solve this problem, it is necessary to calculate the radius of the atoms. To do this, the close packed direction must first be identified. Since the atoms touch each other in this direction, the atomic radius can then be calculated. Looking at the geometry of the crystal, the close packed direction is most likely the [1 0 1] direction.

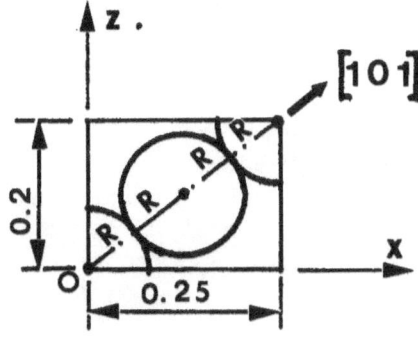

**Fig. E2.4 (d)**

From the figure E2.4 (d), it is evident that:

$$(4R)^2 = 0.25^2 + 0.2^2$$
$$=> R = 0.08 \text{ nm}$$

Going back to the figure showing the (1 0 1) plane, there are an equivalent of two atoms $(1 + 1/4 \times 4)$ in an overall area of 0.3 nm x 0.32 nm. Hence:

$$pd = \frac{Area\_of\_plane\_covered\_by\_atoms}{total\_area\_of\_plane}$$

$$pd_{(101)} = \frac{2\pi R^2}{0.3x0.32} = 0.419$$

Alternatively this may be given as 2 atoms per 0.3 x 0.32 (nm)$^2$ or 2.08 x $10^{19}$ atoms per m$^2$

## 2.5 X-RAY DIFFRACTION

Metal crystal structures are determined by X-Ray diffraction studies. The wavelength of X-Rays is the same order of magnitude as interatomic spacing in crystals. As a result, successive crystal planes diffract X-Rays. Referring to figure 2.8, constructive interference will occur only if the path difference between the rays reflected from the first and second planes is an integral number of wavelengths. From the figure, the path difference = MH' + H'P.

$$MH' = H'P = d \text{ Sin } \theta$$
$$=> \quad 2d \text{ Sin } \theta = n\lambda \qquad\qquad (2.10)$$

where d is the distance between successive planes or the interplanar distance, n is an integer and $\lambda$ is the wavelength of the X-Rays used.

Equation 2.10 is termed Bragg's law. d can then be related to the indices of the plane. For cubic crystals:

$$d = \frac{a}{\sqrt{h^2 + k^2 + l^2}} \qquad\qquad (2.11)$$

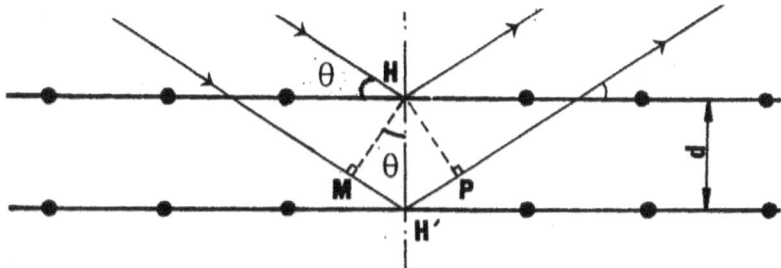

**Fig. 2.8** The diffraction of X-Rays by crystal planes

Several methods are used for X-Ray diffraction in practice. In the powder pattern, monochromatic X-Rays are directed at the material (which is in powder form mixed with a cement and made into a thin filament). The sample is placed at the centre of a circular camera and the diffracted rays are captured on X-Ray film. Several lines representing various diffraction orders (i.e., different values of n in equation 2.10) are obtained. These are then analyzed, but a full description of the procedure is outside the scope of this book. The interested reader is referred to the references listed at the end of the chapter.

## PROBLEMS

2.1 Sketch the unit cell of a crystal of iron and copper at room temperature and calculate the atomic packing factor of each. (Ans: 0.68, 0.74).

2.2 (a) Show the (221), (101) and (110) planes in a simple cubic lattice.
    (b) Explain, giving two examples, what is meant by allotropy.

2.3 Molybdenum has the following characteristics: Lattice constant, a = 0.315 nanometers; crystal structure: BCC; atomic weight: 95.94. If the atomic mass unit is $1.66 \times 10^{-24}$ g, calculate:
    (i) The nearest distance between the centres of two atoms in the unit cell. (Ans: 0.27 nm).
    (ii) The theoretical density in kg/m³. (Ans: 10,223 kg/m³).
    (iii) The distance between successive (110) planes. (Ans: 0.22 nm).
    (iv) The angle between the [111] and [112] directions. (Ans: 19.5°).

2.4 Copper has an FCC crystal structure with a lattice constant a = 0.361 nanometers, and atomic weight 63.54. Calculate the following for copper:

(a) The atomic packing factor of its unit cell. (Ans: 0.74).
(b) The linear density of equivalent sites in the [1 1 1] direction of its crystal. (Ans: $1.6 \times 10^9$ sites/m).
(c) The angle between the [2 1 2] and [1 1 3] directions. (Ans: 25°).
(d) The theoretical density of copper in kg/m³ (atomic mass unit = 1.66 $\times 10^{-24}$ g). (Ans: 8,968 kg/m³).

2.5 Nickel has a face centred cubic crystal structure with an atomic radius, R = 0.1246 nanometers. Calculate the following for Nickel:
(i) The number of atoms per mm² on the (1 1 0) plane of its crystal. (Ans: $1.14 \times 10^{13}$ atoms/mm²).
(ii) The linear density of equivalent sites in the [1 1 0] and the [0 1 1] directions (in sites/m). (Ans: $4.01 \times 10^9$ sites/m for both).
(iii) The Miller indices of the line of intersection of the (1 $\bar{1}$ 0) and (1 1 0) planes. (Ans: [0 0 1]).

2.6 Sketch the unit cell of a hexagonal close packed crystal and calculate the number of atoms/unit cell. Name any three common metals that have this crystal structure. (Ans: 6 atoms/uc).

2.7 Show the (221); (1 0 1) and ($\bar{T}$ 1 0) planes in a simple cubic crystal lattice.

2.8 Calculate the angle of second order diffraction (i.e., n = 2) from the (2 2 1) planes of a BCC lattice of a metal with an atomic radius of 0.12 nanometers when using monochromatic X-Rays with a wavelength of 0.091 nanometers. (Ans: 80°).

2.9 Given that cobalt has a HCP structure:
(i) Name three common metals with a similar structure
(ii) Sketch a unit cell of cobalt clearly showing the lattice constants.
(iii) Determine the number of atoms/unit cell of a cobalt crystal clearly showing how you arrive at your answer.

2.10 Estimate the amount of space not occupied by matter in a 2 m³ block of Nickel at room temperature. If nickel has an atomic radius of 0.1246 nanometers and its atomic weight is 59, estimate the weight of the block. How would your calculated value compare with the actual weight of such a block? In each case, show how you arrive at your answer. (Ans: 0.52m³; 17,900 kg).

2.11 Sketch the unit cells of chromium and nickel and calculate the atomic packing factors of each.

2.12 Given that magnesium has a hexagonal close packed (HCP) crystal structure:
(i) Sketch the unit cell of a crystal of magnesium clearly showing the lattice constants.
(ii) How many atoms are there per unit cell of magnesium? Show how you arrive at your answer. (Ans: 6).
(iii) Assuming perfect packing, calculate the amount of space not occupied by matter in a $1m^3$ block of magnesium at room temperature. (Ans: $0.26 m^3$).
(iv) Name two other common metals, which have a similar crystal structure.

2.13 Calculate the angle for the 1st order diffraction (i.e., n = 1) from the (1 0 0) planes of an FCC crystal lattice of a metal with an atomic radius of 0.11 nanometers when using monochromatic X-Rays with a wavelength of 0.1 nanometers (Ans: 9.2°).

# Further Reading

Anderson, J. C. Leaver, K.D. Rawlings, R.D. and Alexander, J.M, <u>Materials Science,</u> 3rd. ed. ELBS and Van Nostrand Reinhold, Berkshire, 1985.

Callister, W.D. <u>Materials Science and Engineering,</u> 3rd ed. John Wiley and Sons, New York, 1994.

Pascoe, K. J. <u>An Introduction to the Properties of Engineering Materials,</u> 3rd ed. ELBS and Van Nostrand Reinhold, Berkshire, 1982.

Van Vlank, L.H. <u>Elements of Materials Science and Engineering,</u> 6th ed. Addison-Wesley, Reading, 1989.

Cotrell, A. H. <u>An Introduction to Metallurgy,</u> Edward Arnold Ltd. London, 1971.

Schaffer, J. P., et al. <u>The Science and Design of Engineering Materials,</u> Irwin, Chicago, 1995.

# CHAPTER THREE

# CRYSTAL IMPERFECTIONS AND MICROSECTIONS

## 3.1  INTRODUCTION

In real life, crystals depart from the ordered arrangement considered in chapter two. It is these imperfections that dictate most engineering properties of materials (ductility, strength, heat treatability, electrical conductivity, semi-conductor properties, creep strength, etc.). The imperfections are caused by several factors:

1. Imperfect packing during crystallization.

2. Due to thermal vibrations, an atom may jump out of its equilibrium position.

3. During working processes, especially those involving plastic deformation.

4. The presence of impurity atoms.

5. In the case of ionic crystals, by the need to attain a state in which the electrical charges are balanced.

An example is FeO. Since ferrous contains some $Fe^{3+}$ ions, excess $O^{2-}$ are required than suggested by the formula. To balance the charges, some $Fe^{2+}$ sites must be empty. More about this will be said when ceramic crystals are considered in chapter 14.

Some of the defects that may occur are given in the sections that follow.

## 3.2  CRYSTAL DEFECTS

### 3.2.1  Thermal Disorder

Atoms are continuously vibrating about their equilibrium point. The magnitude of vibrations increases with rise in temperature resulting in a net increase in interatomic distance (i.e., a departure from equilibrium spacing). Macroscopically, this manifests itself as thermal expansion.

### 3.2.2  Point Defects

These are zero dimensional interruptions in the regularity of the crystal lattice (i.e., interruptions of 1 atom magnitude in all directions). The main point defects are: (see figure 3.1.)

1. Vacancy: A vacancy occurs when a lattice point, which should normally be occupied by an atom, is empty.

2. Interstitial atom: Also termed self-interstitial atom or interstitialcy. It occurs when an atom similar to the rest occupies a lattice point, which should normally be vacant.

3. Substitutional impurity atom: In this case, an atom of another element occupies a lattice point normally occupied by an atom.

4. Interstitial impurity atom: This case is similar to case 3 above but the impurity atom occupies an interstice.

It is found in practice that all materials have an equilibrium number of defects 1 and 2 at any temperature above absolute zero. This suggests that a lower energy is associated with an imperfect crystal than with a perfect one. Now, the total change in free energy of a system, $\Delta G$, depends on both the change in enthalpy, $\Delta H$, and the change in entropy, $\Delta S$ i.e.:

$$\Delta G = \Delta H - T\Delta S \qquad\qquad (3.1)$$

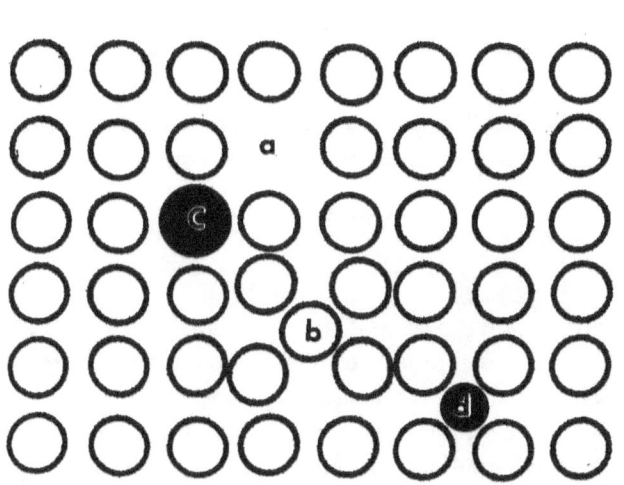

**Fig. 3.1** Point defects: (a) vacancy (b) self interstitial atom (c) substitutional impurity atom (d) interstitial impurity atom.

In order to create one vacancy, energy = $Q_D$ (known as activation energy) has to be supplied to the system to remove an atom from a lattice site to the surface. This increases the enthalpy of the system. But when an atom is so removed, the system becomes disordered, which results in an increase in the entropy. Both effects increase with increase in the number of vacancies as shown in figure 3.2.

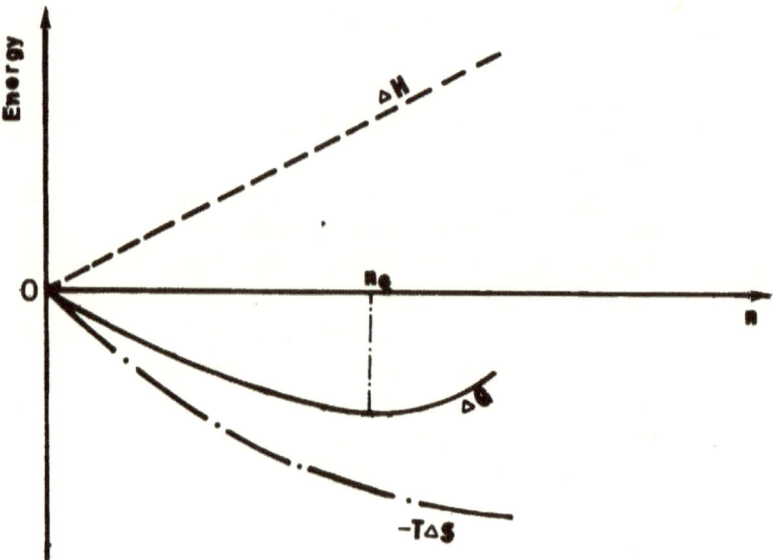

**Fig. 3.2** Schematic representation of the variation of free energy with number of vacancies.

The equilibrium concentration of vacancies, $C_e$ corresponds to the point of minimum free energy and is given by Boltzmann's equation:

$$C_e = \frac{n_e}{N} = A\exp\frac{-Q_D}{kT} \qquad\qquad (3.2)$$

where A is a constant usually taken as 1, $n_e$ is the equilibrium number of vacancies, N is the total number of lattice sites, T is the temperature on the absolute scale and k is Boltzmann's constant.

In metals, the value of $Q_D$ required to create a vacancy is much higher than that required to move an atom into an interstice (the packing is close hence the interstices are small) and hence there are many more vacancies than intestitialcies. In ionic crystals, either may dominate: vacancies in close packed structures and intestitialcies in low packed structures.

Example 3.1: Calculate the number of vacancies in a 1m³ block of copper which is maintained at 1000 °C, given that the activation energy for vacancy formation in copper is 0.9 eV/atom, the atomic weight is 63.5 amu, while the density is 8400 kg/m³. Boltzmann's constant = 8.62 x 10⁻⁵ eV/atom K, and Avogadro's constant = 6.023 x 10²³ atoms/mole.

T = 1000 °C = 1273 K.

Let N = total number of lattice sites in a 1m³ block of copper. From the density, 1m³ block of copper weighs 8400 kg = 8.4 x 10⁶ g, and 63.5 g of copper contains 6.023 x 10²³ atom sites.

$$\Rightarrow N = \frac{6.023 x 10^{23} \, x 8.4 x 10^{6}}{63.5} atom\_sites$$

= 7.967 x 10²⁸ atom sites

Substituting these values into equation 3.2:

$n_e$ = 2.18 x 10²⁵ vacancies

Example 3.2:  The number of vacancies in a metal increases by a factor of five when the temperature is increased from 727 °C to 957 °C. Calculate the activation energy for vacancy formation over this temperature range. Boltzmann's constant = 1.38 x 10⁻²³ J/atom K.

Let   $T_1$ = 727 °C = 1000 K
$T_2$ = 957 °C = 1230 K
$n_1$ = number of vacancies at $T_1$
$n_2$ = number of vacancies at $T_2$

Then from equation 3.2:

$$\frac{n_1}{n_2} = \exp\left(-\frac{Q_D}{k}\left[\frac{1}{T_1} - \frac{1}{T_2}\right]\right) = \frac{1}{5}$$

Substituting the values given above into the equation:

$Q_D$ = 1.19 x 10⁻¹⁹ J/atom

## 3.2.3  Line Defects (Dislocations)

These are one-dimensional (i.e., about one atomic spacing wide but several atomic spacing long) interruptions in the regularity of the lattice. The two common types are the edge and screw dislocations.

Edge dislocation: In this case, an extra half plane of atoms is fitted in the lattice as shown in figure 3.3. This may be considered as due to a slip wave moving along the crystal. The dislocation line, AD in figure 3.3 (b), is the edge of the slipped region i.e., the line separating the slipped from the unslipped portion. An edge dislocation is said to be positive if the extra half plane is above the plane on which the slip is occurring, and negative if it is below this plane. Edge dislocations of the same sign repel while those of opposite signs attract each other.

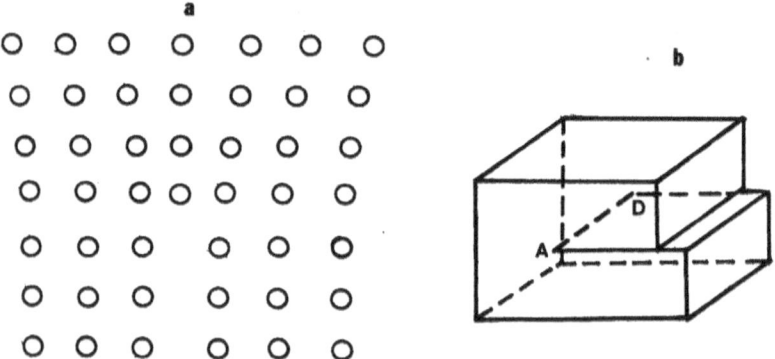

**Fig. 3.3** Edge dislocation: (a) 2-D representation (b) Representation as a slip wave.

Dislocations are identified by their "Burgers vector", **b**, which is the magnitude and direction of the slip resulting from the motion of a dislocation. The Burgers vector can be determined by doing a "Burgers circuit" around the dislocation as follows: (see figure 3.4).

(i) Draw a circuit around the dislocation jumping from atom to atom and count the number of jumps in each direction, until the circuit closes.

(ii) Repeat a similar circuit taking a similar number of jumps in each direction around an unslipped portion of the crystal. This time the circuit does not close. The magnitude and direction of the "step" required to close the circuit is the Burgers Vector.

It may be noted that the Burgers vector of an edge dislocation is perpendicular to the dislocation. This is the distinguishing characteristic of the edge dislocation.

Edge dislocations can glide (i.e., slip in direction of the Burgers Vector) but this takes place only in preferred planes and directions (called slip planes and slip directions respectively). They may also climb (i.e., move in a direction perpendicular to the Burgers vector). This happens if the line of atoms immediately above the dislocation is removed say by diffusion. Climb is always slower than glide since it (climb) involves diffusion of atoms/vacancies.

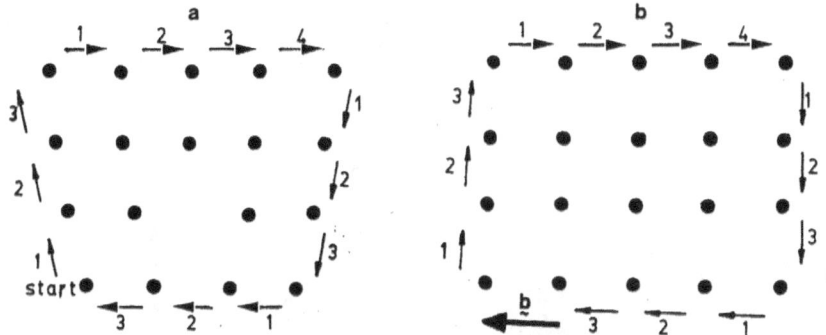

**Fig. 3.4** Illustration of a Burgers circuit for an edge dislocation: (a) circuit around a dislocation (b) circuit around an unslipped part of crystal.

<u>Screw dislocations</u> results from the type of slip shown in figure 3.5. Line AD in the figure is the region between slipped and unslipped regions and is therefore the dislocation. Doing a Burgers circuit as explained previously and shown in figure 3.6, it is noted that the Burgers vector is parallel to the dislocation. Furthermore, the Burgers circuit traces the path of a right hand screw, hence the name screw dislocation.

Screw dislocations also glide but there is no preferred plane hence their glide is easier than those of edge dislocations. No climb is possible with the screw dislocation.

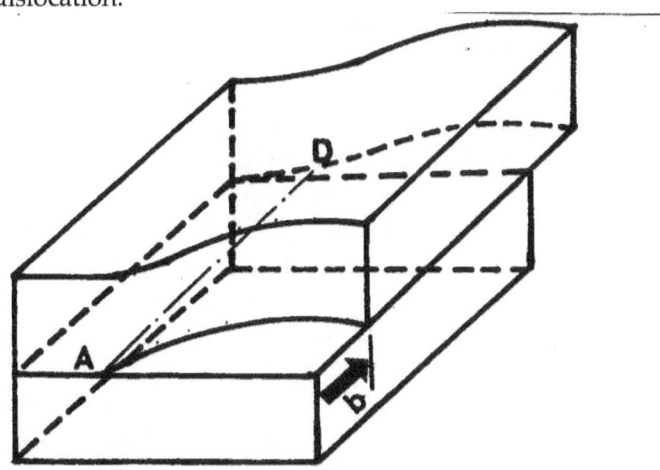

**Fig. 3.5** Schematic of a screw dislocation showing the Burger's vector.

Most dislocations are a mixture of edge and screw dislocations and all are zones of stress (tension on one side and compression on the other) and therefore represent areas of high energy. Changes that take place in the crystal e.g., precipitations therefore start preferentially at dislocations.

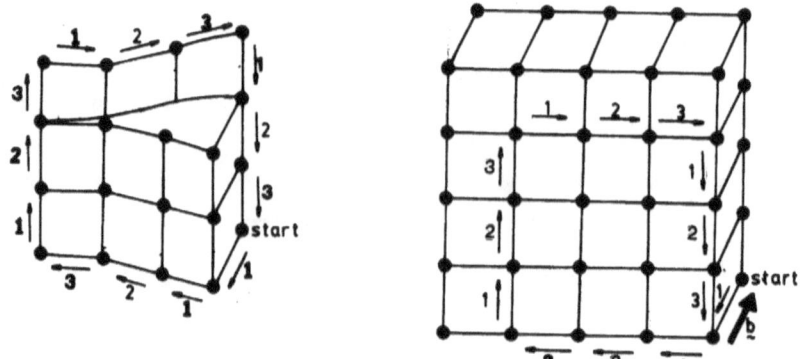

**Fig. 3.6.** Burgers circuit for a screw dislocation

Dislocations can be defined as narrow or wide. The width is defined as that distance over which the displacement of atoms exceeds $b/4$. A dislocation in which $|b|$ = one lattice spacing, a, is termed a unit or perfect dislocation. Dislocations with other widths are said to be imperfect. Those with $|b|$ < a are called partial dislocations while those with $|b|$ > a are termed super dislocations.

The density of dislocations is defined as the total length of dislocations in a unit volume. This is equivalent to the number of dislocations piecing a unit area and is approximately $10^8/cm^2$. It is found from experiment that when a material is plastically deformed, the density of dislocations rises greatly. This happens by a multiplication mechanism termed Frank Reed Source. The source operates as follows: Imagine a dislocation line AB pinned at both ends and acted upon by a stress (figure 3.7) during plastic deformation. The stress pushes the line forcing it to form a loop. Further stressing may cause the loop to fold onto itself and may join up behind the nodes. This results in the formation of a dislocation ring or loop that extends outward while at the same time a new dislocation line is reconstructed. This process is repeated as long as the stress is applied leading to the generation of an unlimited number of dislocation loops from the single source.

### 3.2.4 Planar Defects

These are two-dimensional breaks in the regularity of a lattice and include:

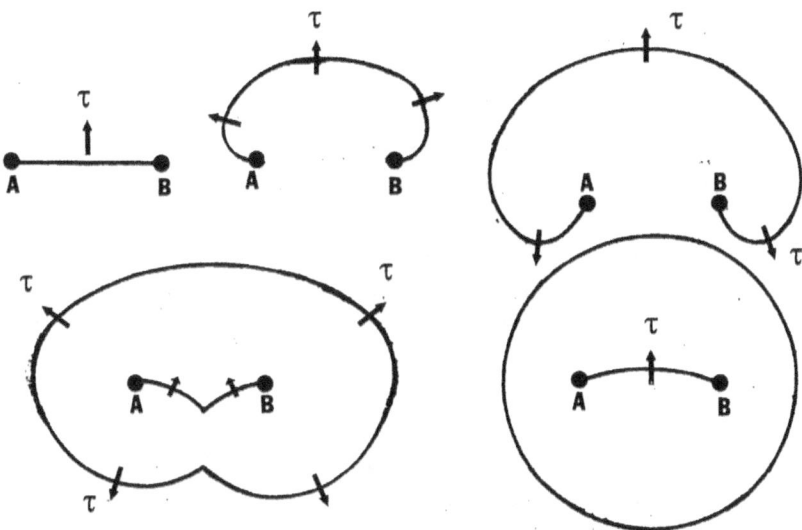

**Fig. 3.7** Generation of dislocation loops from a Frank-Reed source.

Grain boundaries: In polycrystalline materials, each crystal is called a grain. An area of mismatch between two grains, usually approximately 3 atoms thick is the grain boundary (orientation based boundary). Grain boundaries may also be composition based, i.e., the area between phases in multiphase systems or result from differences in the size of crystals. Depending on the angle of orientation between the adjacent grains, grain boundaries may be classified as being small angle or high angle. Small angle grain boundaries (SAGB) are actually a series of aligned dislocations as illustrated in figure 3.8. If the dislocations are edge dislocations, these boundaries are sometimes termed as tilt boundaries. On the other hand, small angle grain boundaries resulting from aligned screw dislocations are termed twist boundaries. Regions of crystals separated by small angle boundaries are sometimes called subgrains.

Grain boundary atoms are less packed than the rest of the crystal and are therefore high energy areas. For this reason, impurity atoms and precipitates tend to form preferentially in these areas. Since systems tend to low energy states, they try to reduce their grain boundary area by grain growth or coarsening (in single phase systems), or by spherodization (in multiphase systems).

Stacking faults: These occur when FCC crystals are plastically deformed. Recall from chapter two that the FCC crystal structure may be obtained by stacking close packed planes of atoms in an ABC ABC ABC

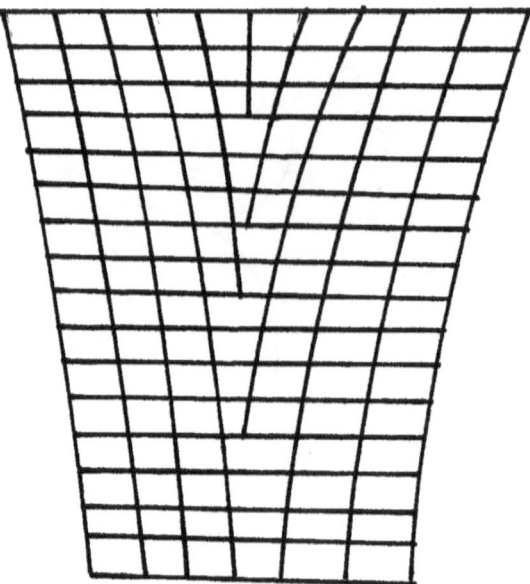

**Fig. 3.8** A series of aligned dislocations constituting a SAGB

stacking sequence. Assume plastic deformation is taking place and that slip has reached the boundary between layer A and layer B. Layer B will have slipped into C position, while the rest of the crystal is unslipped. The ensuing sequence of planes then becomes ABC AC ABC ABC as shown in figure 3.9. This may be rewritten as AB CA CA B... The four planes end up having a HCP type packing (CA CA packing is similar to AB AB packing). The fault in stacking sequence results in a thin HCP layer within an FCC crystal. As with all defects, there is energy associated with a stacking fault that is termed the stacking fault energy (SFE).

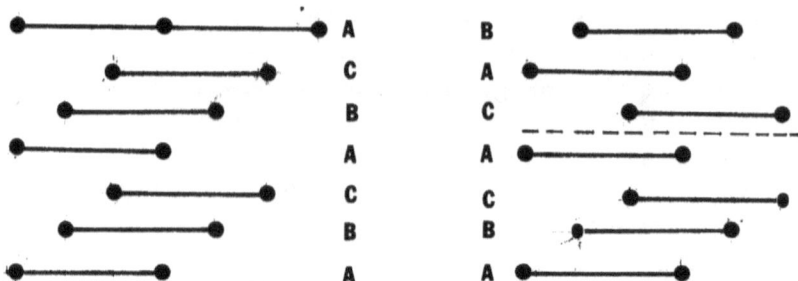

**Fig. 3.9** Schematic representation of a stacking fault

<u>Twins</u>: A crystal is said to be twinned if one part is a mirror image of the other. The plane of symmetry is termed the twin plane (see figure 3.10). Twins may be produced by mechanical deformation (mechanical twins) or by annealing (annealing twins). Mechanical twins occur in HCP and BCC crystals as a result of impact loading. It takes place on specific planes (twin planes) in specific directions. They occur when slip is restricted.

<u>Surfaces</u>: the atoms on the surface do not have neighbours on one side and hence represent a break in the lattice continuity. Effects like surface tension and adsorption exemplify their higher energy.

### 3.2.5 Bulk Defects

These are breaks in the regularity of the lattice that are three-dimensional. They include pores, slug inclusions, etc. that may be introduced in materials by the method of manufacture e.g., casting, welding, etc.

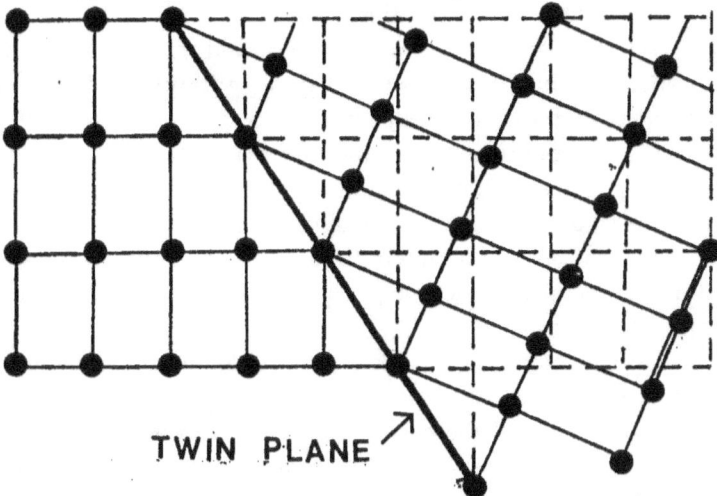

TWIN PLANE

**Fig. 3.10**   Illustration of a twinned crystal

## 3.3 MICROSTRUCTURES
### 3.3.1 Introduction:

A microstructure refers to the structure and arrangement of several grains and is characterized for each material by the grain size, shape, and orientation. The size of grains may vary from less than microscopic to several centimeters. Microstructures have a direct bearing on material properties hence their importance.

## 3.3.2   Preparation of Microsections

Before a microstructure can be observed under the microscope, it has to be prepared. The stages in the preparation are: sectioning, mounting, identification, grinding, polishing, cleaning and etching. Care is needed in all stages to ensure that the preparation process produces no changes in microstructure. In particular, all processes generating heat and/or pressure are to be avoided or controlled.

Sectioning: A portion, which is representative of the material, is selected and sectioned off. If methods likely to cause recrystallization, tempering, etc., (e.g., gas cutting, sawing or grinding) are used, the cut is made far enough. Usually, the final sectioning is done using abrasive wheels (bonded $Al_2O_3$ or SiC) under running water.

Mounting and identification: Small or irregular sections are mounted in resin for ease of handling (also called plastic embedding). The mounted specimens are then engraved for identification.

Grinding: The purpose is to remove the deformed material and render the surface reflective and plane. Usually, coarse grinding is done using abrasive mounted on paper or cloth and attached to a rotating wheel. Running water is continuously passed over the wheel for cooling and for removing loose particles. Fine grinding, again under running water, follows on successively finer grains of silicon carbide paper. To avoid directional effects, rotate the specimen 90° after every stage. To wash off particles from the coarser grades of paper, the specimen is held under running water after every stage.

Polishing (Mechanical): This is done, to produce a mirror finish on the specimen, using a special cloth attached to a rotating wheel. Abrasive, in the form of powder or paste (diamond dust, $Al_2O_3$, chromic oxide, etc.) is used on the moist cloth and the specimen is rotated opposite to the rotation of the wheel to avoid directional effects.

Polishing may also be done electrolytically (anodic polishing) or chemically. After polishing, the specimen is dried rapidly by rinsing in alcohol (which will dissolve any water including those in the pores/scratches) then blowing with hot air.

Etching is done to impact different appearances to different constituents of the microsection. It involves selective chemical corroding to produce shades of varying degrees. The specimen is immersed in the etching solution for a given time (important) and then washed in running water, rinsed in alcohol, and dried in hot air. For cast iron and steel, the usual etching agents are nital (2% solution of nitric acid in alcohol) and solution of picric acid in alcohol.

The method of preparation described here involves manual processes. In current practice, most of the procedures are automated.

### 3.3.3 The Metallurgical Microscope

Prepared micro sections are observed in the metallurgical microscope whose distinguishing characteristic is that it employs the principle of reflected light. In using the microscopes, start from the lowest magnification and progress to the highest.

For long life of the microscope, the optical system must be kept free of dust and the lenses should not be touched.

### 3.3.4 Macro-Examination and Sulphur Printing

Macro-examination refers to examination with naked eye or under low power of the microscope. It is useful in failure analysis (determining type of failure e.g., fatigue); determination of manufacturing defects (slag inclusions, blow holes, lack of fusion, non-uniform segregation, etc.); physical defects (laminations, pipes).

Sulphur printing reveals the presence of sulphur (an impurity) in steels. Sulphur reacts with acid to evolve hydrogen, which affects bromide (photographic) paper. The photographic paper is soaked in dilute $H_2SO_4$ and placed on the selected surface. If the steel contains sulphur, hydrogen will be evolved and the photographic paper will be affected.

## PROBLEMS

3.1 Distinguish between edge and screw dislocations in a crystal lattice.

3.2 What are the 4 main types of point defects in crystals and of what importance are they (the defects) to the properties of a material?

3.3 Figure Q3.3 shows a hypothetical two-dimensional crystal lattice. Each symbol represents an atom. Different symbols mean different elements. Identify each crystal imperfection present in the diagram by giving the letter and number that identify its position. (Ans: Vacancy, C3; self-interstitial, C2; edge dislocation, B1; interstitial impurity, A3; substitutional impurity, B3).

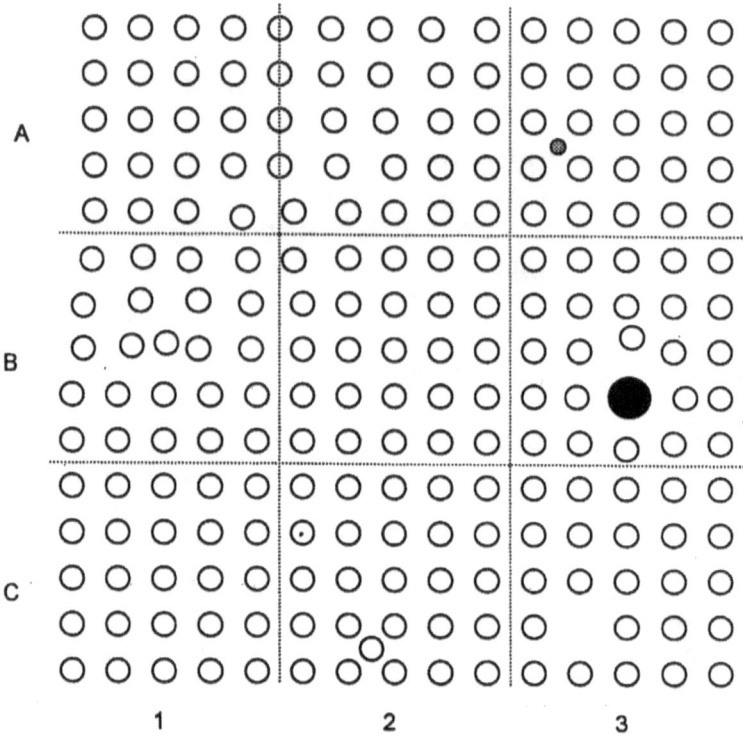

1          2          3

**Fig. Q3.3**

3.4 Give a brief account of the technique and purpose of sulphur printing.

3.5 Distinguish clearly, with the aid of sketches, between edge and screw dislocations.

3.6 Give a brief account of the procedure to be followed in preparing micro sections mentioning any precautions to be taken and why.

3.7 It is observed that all materials have an equilibrium concentration of point defects at any temperature above 0 K. How can you explain this?

3.8 Explain, using a sketch of an edge dislocation, how a "Burgers circuit" is done. In what ways would a screw dislocation differ from the edge dislocation you have illustrated?

3.9 Prove, by performing a Burgers circuit, that the Burgers vector of a screw dislocation is parallel to the dislocation.

3.10 Calculate the activation energy for vacancy formation in silver if the equilibrium number of vacancies at 800 °C is 3.6 x 10²³/m³. Other constants for silver are: atomic weight, 107.9 amu, density, 9 500 kg/m³. Also, Avogadro's number = 6.023 x 10²³ atoms/mole, and Boltzmann's constant = 1.38 x 10⁻²³ J/atom K. (Ans: 1.8 x 10⁻¹⁹ J/atom).

3.11 Calculate the number of vacancies in silver at 500 °C if the number of vacancies at 900 °C is 10²⁴/m³. The activation energy for vacancy formation is 1.1 eV/atom. (Ans: 3.59 x 10²¹/m³).

## Further Reading

Anderson, J. C. Leaver, K.D. Rawlings, R.D. and Alexander, J.M. Materials Science, 3rd. ed. ELBS and Van Nostrand Reinhold, Berkshire, 1985.

Dieter, G. Mechanical Metallurgy, 3rd. Ed. McGraw-Hill, New York, 1986.

Vander Voort, G. F. Metallography, Principles and Practice, McGraw-Hill, New York, 1984.

# CHAPTER FOUR

# TESTING OF MECHANICAL PROPERTIES

## 4.1  INTRODUCTION

The mechanical properties of material are those properties, which define the response of a material to the application of forces. Determination of mechanical properties is useful for selection of materials, design and for quality control.

## 4.2  THE TENSILE TEST

### 4.2.1  Stress and Strain

Consider an arbitrary body in equilibrium under action of a system of external forces. The external forces induce internal forces, which are transmitted across any cross section. The intensity and direction of this force varies from point to point and if on an area, $\Delta A$, the force is $\Delta P$, the stress, $\sigma$, is defined as:

$$\sigma = \lim_{\Delta A \to 0} \frac{\Delta P}{\Delta A} = \frac{dP}{dA} \qquad (4.1)$$

Stress is a vector and can be resolved into 3 components. Two of these are tangential while one is normal to the surface of the section. The normal component of stress is termed the normal or direct stress, $\sigma$, and may be tensile (away from the surface) or compressive (towards the surface). The tangential components are termed shear stresses, and are designated by $\tau$. The SI unit of stress is $N/m^2$ or Pascal, Pa. To have manageable figures, the units $MN/m^2$ (MPa) and $N/mm^2$ are also used.

If (and only if) the force system is a single uniaxial force, P, and the cross section area of the material perpendicular to the force is uniform and constant (=A), then $\sigma = P/A$.

The application of a force also causes an extension, $\Delta L$ in the material. If this occurs over a length, $L_o$, (termed the gauge length), then the engineering strain, $\varepsilon$, may be defined as:

$$\varepsilon = \frac{\Delta L}{L_o} \qquad (4.2)$$

It is also possible to define true, logarithmic or natural strain, $\varepsilon_n$, (which refers to the instantaneous length) as:

$$\varepsilon_n = \int_{L}^{L_f} \frac{dL}{L} = \ln\left[\frac{L_f}{L_o}\right] \qquad (4.3)$$

where $L_f$ is the final length, and ln stands for the natural base of logarithms. Now,

$$\varepsilon = \frac{L_f - L_o}{L_o} = \frac{L_f}{L_o} - 1$$

$$\Rightarrow \frac{L_f}{L_o} = 1 + \varepsilon$$

$$\Rightarrow \varepsilon_n = \ln\left[\frac{L_f}{L_o}\right] = \ln(1 + \varepsilon) \qquad (4.4)$$

At low strains, the natural and engineering strains are approximately equal.

An extension in the longitudinal direction is accompanied by a contraction in the lateral direction and vice versa. Within the elastic range, the ratio between the longitudinal strain and the lateral strain (for round specimens, this is the change in diameter divided by the original diameter) is a constant. The constant of proportionality is termed Poisson's ratio, and may be defined as:

$$\nu = -\frac{lateral\_strain}{longitudinal\_strain} \qquad (4.5)$$

Poisson's ratio is a property of the material and is about 0.3 for most metallic alloys.

## 4.2.2   Performance of the Tensile Test

This test is used to determine several mechanical properties of metallic alloys and is probably the most often performed mechanical test. In Kenya, it is performed in accordance with the specifications of the Kenya Standard KS 06-141: Specification for the Tensile Test. A standardized specimen, generally of the form shown in figure 4.1, is loaded in uniaxial tension in a tensile testing machine. The load is recorded against the elongation of a specified original gauge length, $L_o$. From the predetermined dimensions of the specimen, the engineering stresses are determined and plotted against the engineering strains. For mild steel, the general form of the stress- strain curve is shown in figure 4.2.

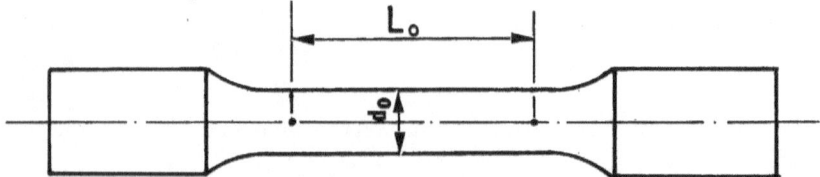

**Fig. 4.1** One form of the tensile test specimen. $L_o$ = the original gauge length, $d_o$ = original diameter.

Among the properties which can be determined from the tensile test are:

      1.   The limit of proportionality: This corresponds to point A. Below this point stress is proportional to strain.

      2.   The elastic limit: corresponding to point A'. Below this point, all the strain is recoverable i.e., the specimen returns to its original length on removal of the force.

      3.   The yield stress is the stress corresponding to point Y, the first point at which an increase of strain occurs at constant (or reduced) stress.

      4.   The lower yield stress is the stress corresponding to point Y', the lowest point to which the stress reduces after yielding.

      5.   The ultimate tensile strength is the stress corresponding to the point of maximum force, U.

      6.   The breaking stress is the stress at the point of breaking, B. This quantity is not a material property since it depends on other parameters like the speed of loading, etc.

      7.   The ductility of the material is measured by two parameters: the amount of plastic strain after fracture, $\delta$, and the percent reduction in area, $\psi = (A_o - A_f)/A_o * 100$ %, where $A_o$ is the area before the test and $A_f$ is the smallest area after fracture. The ductility of a material is a measure of the amount of plastic deformation it can take before breaking.

After necking, most of the extension will take place around the necked region (see references at the end of the chapter). The elongation after fracture, $\Delta L$ therefore consists of two parts: one part due to necking which is not proportional to the gauge length, and one part proportional to the gauge length, i.e.:

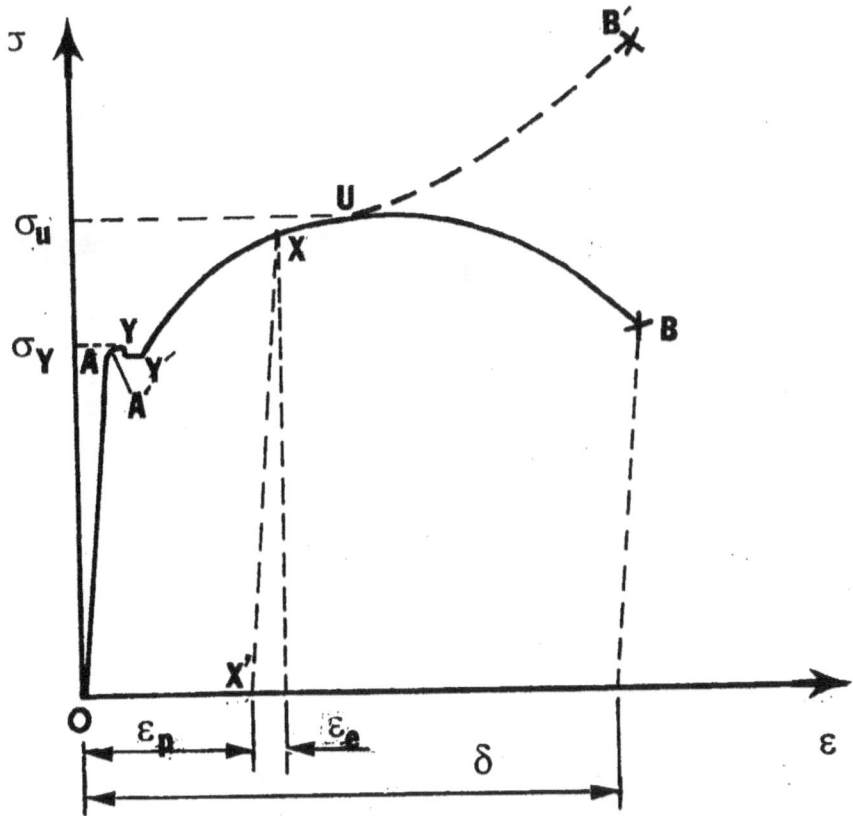

**Fig. 4.2** The general form of the stress-strain curve for mild steel.

$$\Delta L = a + b \times L_o \qquad (4.6)$$

where a and b are constants. Experiments have shown that a is proportional to $A^{1/2}$ i.e:

$$a = C A^{1/2} \qquad (4.7)$$

Hence,
$$\Delta L = C A^{1/2} + b \times L_o \qquad (4.8)$$

C and b are properties of the material and are termed Urwin's constants. Then:

$$\delta = \frac{\Delta L}{L_o} x100\% = \left[ \frac{C\sqrt{A}}{L_o} + b \right] x100\% \qquad (4.9)$$

The implication of equation 4.9 is that only geometrically similar test pieces of the same material (i.e., test pieces with the same $A/L_o$ ratio) will give the same value of percent elongation. This statement is termed Barba's Law.

Most standards specify a $L_o/A$ ratio of 5 or 10. Results from non-proportional test pieces may be converted to standard values by using equation 4.9.

For plotting the true stress-strain graph in metallic alloys, the principle of constancy of volume is used up to point U. (Note: the stress strain curves obtained when ceramics and plastics are tested will be considered in more detail in chapters 13 and 14.) Hence, before necking starts, extension is uniform throughout the gauge length, and:

$$A_o L_o = AL \qquad (4.10)$$

So the true stress, $\sigma_t$, is given by:

$$\sigma_t = \frac{P}{A} = \frac{PL}{A_o L_o} \qquad (4.11)$$

After start of necking, it is best to determine the actual diameter of the neck. The true strain is calculated from equation 4.4.

At any point past the elastic limit, the strain consists of an elastic component, which is recoverable on removal of the load, and a plastic component, which is permanent. As a result, if the test specimen is unloaded at point X, the unloading curve will be XX'. When reloaded, the material will not yield until point X. In other words, straining has increased the yield stress. The material is said to be strain hardened. The rate of strain hardening is measured by the static strain hardening exponent, n.

The engineering stress-strain curve, which follows UB in figure 4.2, is obtained when stress is calculated on the basis of the original cross section area. After point U however, the material begins to "neck". The true stress-true strain curve, UB', is plotted on the basis of the actual cross section area. The stress corresponding to B' is the true fracture stress.

High strength steels and non-ferrous metals and alloys do not show a distinct yield point. Instead, a proof stress, at a given plastic strain, x, (usually 0.2 %) is defined (see figure 4.3) and used in place of the yield stress. In other words, the 0.2 % proof stress is the stress required to cause a permanent strain of 0.2 %. Other stress-strain curves that can be obtained are shown in figure 4.4.

Below point A in figure 4.2, the stress is proportional to the strain. This phenomenon is termed Hooke's Law and the proportionality constant, E is called Young's modulus or modulus of elasticity, i.e:

$$\sigma = E\varepsilon \qquad\qquad\qquad (4.12)$$

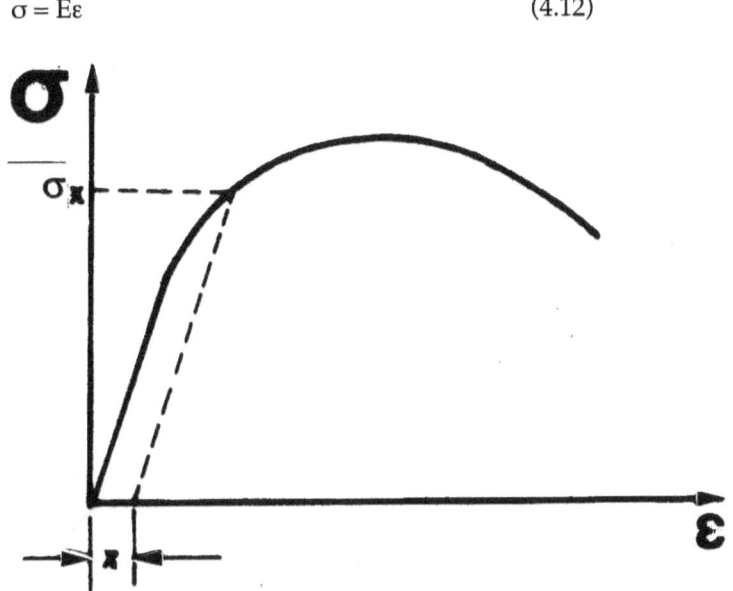

**Fig. 4.3** Definition of proof stress

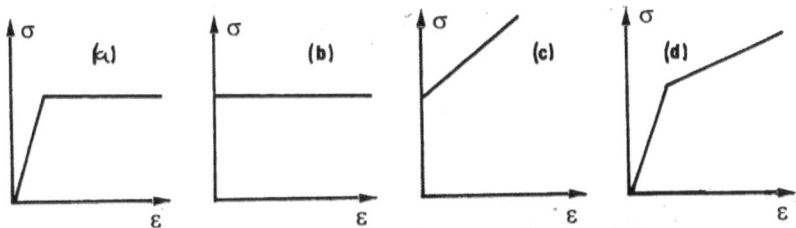

**Fig. 4.4** Idealized forms of stress-strain curves: (a) elastic--perfectly plastic (b) rigid--perfectly plastic (c) rigid--linear strain hardening (d) elastic--linear strain hardening.

E is another important material property and for steels is approximately 200 GPa. A similar relationship exists between the shear stress, $\tau$, and the shear strain, $\gamma$, i.e:

$$\tau = G\gamma \qquad (4.13)$$

The constant G is the modulus of rigidity or shear modulus and is approximately 80 GPa for steels.

When a material is stretched, work is done since a force moves. In the elastic range, this work is stored in the material as strain energy and may be recovered when the force is released. The ability of a material to store elastic energy is termed its resilience. A measure of a material's resilience is its modulus of resilience, $U_R$. This is the maximum amount of energy the material can store as strain energy per unit volume of the material. Consider a specimen, gauge length, $L_o$, cross section area, A, which undergoes an extension, $\Delta L$ (figure 4.5):

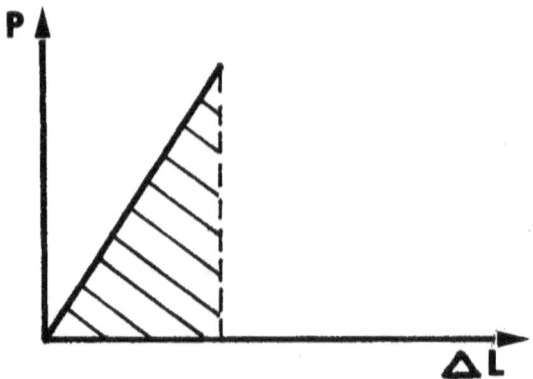

**Fig. 4.5** Calculation of modulus of resilience

Work done = Area under load displacement curve
$\qquad = 1/2\,P \times \Delta L$
But, $P = \sigma \times A$
=> Work done $= 1/2\,\sigma \times A \times \Delta L$

$$\therefore U_R = \frac{\sigma A \Delta L}{2\,AL} = \frac{1}{2}\sigma\frac{\Delta L}{L_o} = \tfrac{1}{2}\sigma\varepsilon \qquad (4.14)$$

Since maximum energy is stored when the stress equals the elastic limit $\sigma_e$, the corresponding strain being equal the elastic strain, $\varepsilon_e$:

$$U_R = \frac{1}{2}\sigma_e \varepsilon_e$$

But, $\varepsilon_e = \dfrac{\sigma_e}{E}$

Therefore,   $U_R = \dfrac{\sigma_e^{\,2}}{2E}$                    (4.15)

For most materials, the elastic limit may be replaced by the yield or proof stress without introduction of appreciable error. From the above, it is clear that the modulus of resilience is simply the area under the stress-strain curve up to the elastic limit.

In the plastic range, the work done is absorbed in the plastic deformation. The toughness of a material defines its ability to absorb energy before fracture. The modulus of toughness, $U_T$, is the energy per unit volume before fracture, and may be estimated from:

$$U_T = \frac{1}{2}(\sigma_Y + \sigma_{UTS})x\delta$$                    (4.16)

where the subscripts Y and UTS refer to the yield or proof stress and the ultimate tensile strength, respectively.

Example 4.1: Figure E4.1 shows the low strain portion of a stress strain curve obtained from a tensile test. Other readings were: initial diameter, 12.8 mm; initial gauge length, 50.8 mm; final diameter, 10.06 mm; length after fracture, 59.18 mm; maximum load, 47.4 kN; load at fracture, 36.5 kN. Determine: (a) the limit of proportionality (b) 0.2 % proof stress (c) Young's modulus (d) modulus of resilience (e) % elongation at fracture (f) % reduction in area (g) the true fracture stress.

Answers:

(a) This is the stress corresponding to the point of departure from linearity. From the figure, the value is 235 MPa.

(b) Draw a line parallel to the straight portion of the curve and passing through 0.2 % point on the strain axis. Read the stress corresponding to the point where this line crosses the curve: 280 MPa.

(c) E = slope of the linear portion = $^{160\,MPa}/_{0.25\,\%}$ = 64 GPa

(d) The strain corresponding to the 0.2 % proof stress is 0.66 % (obtained by drawing a vertical line from 280 MPa, and reading its intercept on the strain axis).

$U_R = \,^{1}/_{2}$x $\sigma_Y$ x $\varepsilon_Y = \,^{1}/_{2}$ x 280 x 0.0066 MPa = 0.9 MPa (or J/m³).

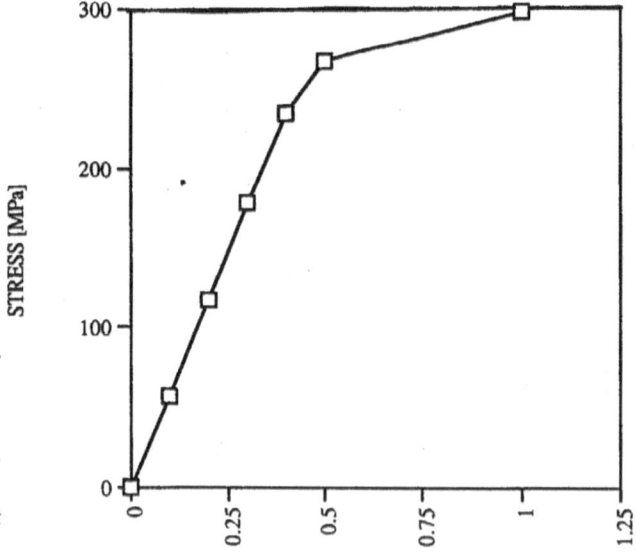

**Fig. E4.1**

(e) % Elongation at fracture can be calculated using the initial and final lengths as:

$$\delta = \frac{59.18 - 50.8}{50.8} x100\% = 16\%$$

(f) % Reduction in area can be calculated using the initial and final diameters:

$$\psi = \frac{A_o - A_f}{A_o} = \frac{d_o^2 - d_f^2}{d_o^2} = \frac{12.8^2 - 10.06^2}{12.8^2} x100\_\% = 38\_\%$$

(g) True fracture stress is calculated using the final diameter and load at fracture:

$$\sigma_t = \frac{4x36.5x10^3}{\pi(1.006\ )^2 x10^{-4}} = 460MPa$$

Example 4.2: Figure E4.2 shows the load extension graph obtained from a tensile test of a stainless steel specimen. The insert is an expansion of the low strain portion. The initial diameter was 12.8 mm while the initial gauge length was 50.80 mm. Determine: (a) the 0.2 % proof stress; (b) Young's modulus; (c) the modulus of resilience; (d) the value of Poisson's ratio if the diameter was 12.791 mm when the load was 60 kN; (e) the % elongation at fracture.

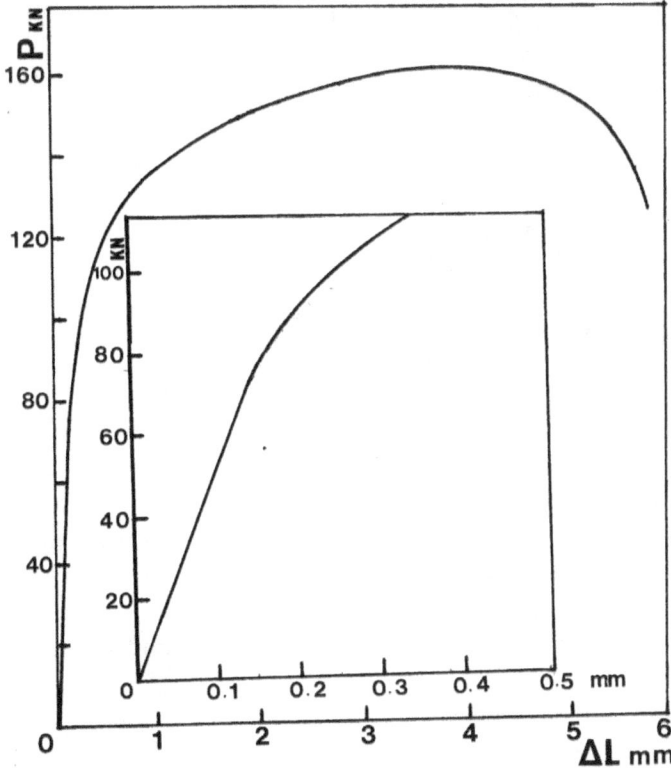

Fig. E4.2

(a) The elongation corresponding to a strain of 0.2 % is first determined:

$\varepsilon = 0.2\% = \Delta L / L_o$  => $\Delta L = 0.10$ mm

The load that produces a plastic elongation of 0.1 mm is then determined from the insert graph (draw a line parallel to the straight potion and passing through 0.1 mm on strain axis) = 109 kN. Then:

$$\sigma_{0.2} = \frac{P_{0.2}}{A_o} = \frac{4 x 109 x 10^3}{\pi (12.8 x 10^{-3})^2} = 847 MPa$$

(b) Taking corresponding values of P and ΔL in the linear portion of the insert graph (60 kN, and 0.115 mm):

$$E = \frac{\sigma}{\varepsilon} = \frac{PL_o}{A_o \Delta L} = \frac{4x60,000x50.8}{\pi(12.8x10^{-3})^2 x0.115} = 206\_GPa$$

(c) $U_R$ :

$$U_R = \frac{\sigma_{0.2}^2}{2E} = 1.74J/m^3$$

(d) At 60 kN, ΔL = 0.115 mm:

$$v = -\frac{\Delta d}{d_o} x \frac{L_o}{\Delta L} = \frac{12.8 - 12.79}{12.8} x \frac{50.8}{0.115} = 0.3$$

(e) Maximum elongation = 5.8 mm (read from graph). Hence:
δ = 5.8 / 50.8 x 100 % = 11 %

## 4.3  THE HARDNESS TESTS

The hardness of a material determines its resistance to penetration by other materials. In the standardized tests, a standard indenter is pushed into the material with a given force and the size of the residual impression measured. For rubbers, hardness is measured by the rebound of a ball from the surface.

Hardness tests are quicker and cheaper to perform than tensile tests and the results so obtained may help estimate the tensile strength. The three main hardness scales in common use for metals are:

### 4.3.1  Brinell Hardness Test

The indenters are hardened and polished steel balls of diameters, D = 10 mm, 5 mm, 2.5 mm and 1 mm. These are pressed onto the surface for 10-30 seconds, with forces varying from 3000 kg to 1.25 kg (29.4 KN to 12.25 N). After removal of the load, the surface diameter of the residual impression is measured. The hardness is determined as: (see Kenya Standard KS 06 - 873: Method for the Brinell Hardness Test).

Hardness = load/(Area of impression)
Referring to figure 4.6:

$$x = \sqrt{(\frac{D}{2})^2 - (\frac{d}{2})^2} = \frac{1}{2}\sqrt{(D^2 - d^2)} \qquad (4.17)$$

$$t = D/2 - x = 1/2[D - (D^2 - d^2)^{1/2}] \qquad (4.18)$$

Area of impression $= \pi Dt = \pi D/2[D - (D^2 - d^2)^{1/2}] \qquad (4.19)$

Then the Brinell hardness, BHN, is given by:

$$BHN = \frac{P}{\frac{\pi D}{2}[D - \sqrt{(D^2 - d^2)}]} \qquad (4.20)$$

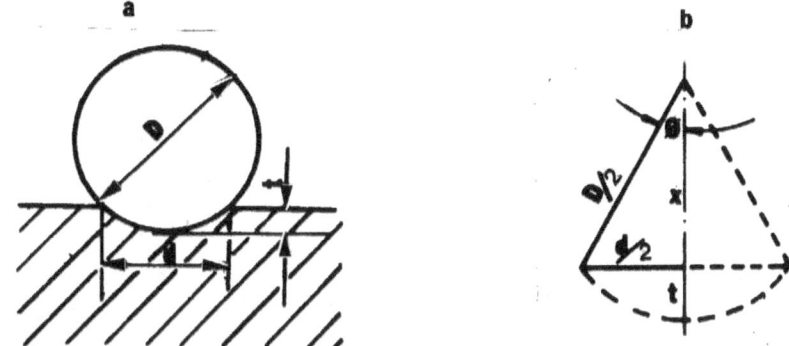

**Fig. 4.6** The Brinell Hardness Test (a) geometry of ball penetration (b) Relation between depth of penetration and included angle

Usually, values of BHN are tabulated as a function of d and can therefore be read directly from the relevant charts. Such charts/tables form part of the standards. The Brinell hardness is expressed as: x HB D/P/t where, x = hardness in kg/mm²; HB = Brinell hardness; D = diameter of the ball used; P = load used in kg; and t = indentation time in seconds, e.g., 210 HB 5/750/15. If P = 3000 kg, D = 10 mm, t = 10-15 s, these may be omitted and the hardness expressed simply as 210 HB.

For reproducibility and validity of results, the impressions must be geometrically similar. Hence, when the load is changed, the ball diameter must also be changed. Geometrical similarity is achieved if the included angle $\phi$ (figure 4.6) is constant. Now:

$$x = \frac{D}{2}\cos\phi \qquad\qquad (4.21)$$

$$t = \frac{D}{2} - x = \frac{D}{2}(1-\cos\phi)$$

$$BHN = \frac{P}{\pi Dt} = \frac{2P}{\pi D^2(1-\cos\phi)} = \frac{P}{D^2} \cdot \frac{2}{\pi(1-\cos\phi)} \qquad (4.22)$$

In order for both BHN and cos $\phi$ to be constant, $P/D^2$ must be constant. This constant is called the Load Ratio and the following values of load ratio are recommended: 30 (ferrous materials); 10 (aluminium and copper alloys); 5 (pure aluminium); 2.5 (bearing materials) and 1.25 (lead, tin).

For valid results, no deformation should be visible on the underside of the specimen which can be ensured if the thickness of the specimen is at least 10 times the penetration depth and d = 0.2 to 0.5 D.

Hardness values higher than 450 HB should not be determined with the steel ball since the indenter then undergoes appreciable deformation and the recorded hardness is lower than the actual value. For such hard materials, a tungsten carbide indenter (E = 720 GPa) or another hardness scale, should be used.

Advantages of Brinell Scale

1. The impression diameter is large, hence the method is insensitive to slight errors in measurement of d.

2. The large impression covers several grains hence the measured hardness is a good average (especially suitable for large grained castings and forgings).

Disadvantages

1. The need to observe correct load ratio and the varying indenter diameters leads to a rather complex choice of loads and indenter diameters.

2. Cannot be used to test hard materials for the reasons given above.

3. Cannot be used to test small (critical) areas.

Example 4.3: A Brinell hardness test was performed on a 15 mm thick piece of alloy cast iron using a load of 750 kg and a 5 mm diameter steel ball. An impression diameter of 1.29 mm was measured. Determine if the value of hardness obtained is valid by considering at least three criteria. If not valid, what steps would you take to obtain a valid result?

Answer:

The three criteria to check are the load ratio, the thickness of the specimen, and the suitability of a steel ball for the test.

(1)   $P/D^2 = 750/5^2 = 30$

Since the correct load ratio for cast iron is 30, the choice of load ratio is correct.

(2) For correct specimen thickness, d = 0.2 to 0.5 D and specimen thickness should be at least 10 times the penetration depth.

Allowable range of d = 1 mm to 2.5 mm. Since d = 1.29 mm, this is satisfactory.

Penetration depth $= D/2\,(1\text{-cos }\phi)$

And $\phi = \sin^{-1}(d/D) = 15°$   (see figure 4.6)

=> Depth of penetration = 2.5 x (1-cos 15°) = 0.08 mm.

Thickness = 15 mm which is adequate.

(3) Applying equation 4. 17 and the given values of d = 1.29 mm and D = 5 mm:
BHN = 564 HB

This is higher than the 450 HB allowed for a steel ball. Hence, the value obtained is not valid. The remedy for this is to use either a tungsten carbide ball or another scale of hardness testing e.g., Vickers.

## 4.3.2   The Vickers Hardness Test

In this case, the indenter is a square based diamond pyramid with an included angle of 136° (Kenya Standard KS 06-70:1 Specification for Vickers Hardness Test). After indentation, the diagonals of the impression left are measured (figure 4.7). The Vickers hardness = load/Area of impression. From figure 4.7, area of impression = 4 x Area CEF = 4 x 1/2 CE x DF.

But, $CE^2 + BC^2 = 2CE^2 = d^2 => CE = d/2^{1/2}$      (4.23)

$$DF = \frac{AD}{2} / \sin(\theta/2) = \frac{d}{(2\sqrt{2})\sin(\theta/2)} \qquad (4.24)$$

=> Area of the impression is:

$$4x\frac{1}{2}x\frac{d}{\sqrt{2}}x\frac{d}{2\sqrt{2}\sin(\theta/2)} = \frac{d^2}{2\sin(\theta/2)} \qquad (4.25)$$

Then the Vickers Hardness, HV, is:

$$\frac{Px2\sin(\theta/2)}{d^2} = \frac{1.854P}{d^2} kg/ mm^2 \qquad (4.26)$$

(Since $\theta = 136°$).

Tables based on equation 4.26, are available for converting d directly to hardness units. The measured hardness is expressed as: x HV P/t, where x = hardness in kg/mm², HV = Vickers hardness, P = load used, t = time in seconds e.g., 300 HV 30/10. If t=10-15 seconds, this may be omitted and the hardness expressed simply as 300 HV 30.

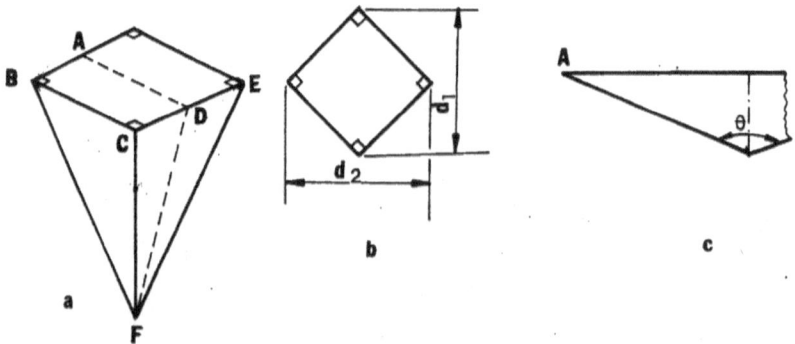

**Fig. 4.7** The Vickers hardness test (a) 3--D geometry of indentation (b) Surface of the indentation (c) tip of the indentation.

In older literature, expressions such as VHN (Vickers hardness number), DPH (diamond pyramid hardness), VPH (Vickers Pyramid hardness) and VPN (Vickers Pyramid number) may be found.

For valid results, the thickness of the test piece should be at least 1.5 d.

Example 4.4: The Vickers hardness of a surgical blade having a thickness of 0.2 mm was determined as 680 HV10. Check if the blade thickness was sufficient for the test.

Answer:  $HV = 1.854 \, P/d^2$

From the way the hardness is expressed, $HV = 680 \, kg/mm^2$, and load = 10 kg. Hence, $d = 0.165 \, mm$

The minimum allowable thickness is $1.5d = 0.25 \, mm$. Since the blade thickness was 0.2 mm, the specimen was too thin and the reading obtained is invalid.

Advantages of the Vickers scale:
1.  It gives a continuous scale of hardness with the same indenter for all hardness ranges.
2.  The hardness determined is independent of the load since all indentations are geometrically similar.
3.  Indentations are small (approximately 0.2 mm) allowing for testing of:
    a.  Thin samples.
    b.  Small critical areas.
    c.  Surface hardened parts.

Furthermore, the tested component is not seriously affected and may be used after test.
4.  May be used for micro hardness (measurements on specific phases in a microstructure) measurements.

Disadvantages:
1)  Due to the small indentations, the hardness is affected by heterogeneities in the material. It is therefore unsuitable for large grained structures.
2)  Due to the small indentations, a small error in measurement of d will lead to large errors in the hardness.
3)  Requires proper surface preparation.

## 4.3.3 Rockwell Hardness Test

The indenter in the Rockwell test is a diamond cone with an angle of 120° and a tip radius of 0.2 mm, or steel balls of various diameters (due to its American origin, the ball diameters are expressed in inches). The Kenya Standard for this test is KS 06 - 71: Method for the Rockwell hardness test. Unlike the Brinell and Vickers hardness test (where the area of indentation is measured), the depth of penetration is measured and converted to arbitrary hardness units in the Rockwell scale.

There are several Rockwell scales available, which are achieved by varying the load and indenter. The main ones are Rockwell C (indenter is diamond cone, load 150 kg) for hardened steel/cast iron parts; Rockwell B (indenter is $1/16$" steel ball and load of 100 kg) for soft steel and non- ferrous alloys; and Rockwell A (same indenter as C, load 60 kg) for carbides and case hardened parts). Many more scales (Rockwell D through to Rockwell Y) are available for specific applications like superficial hardness tests, rubbers, bonded materials, etc.

The procedure for the Rockwell hardness test is:

1) A pre-load (10 kg for Rockwell A, B and C) is applied. The penetration, $t_o$, under the preload acts as the reference point.
2) The main load (140 kg for C; 50 kg for A and 90 kg for B) is applied. The penetration depth from the reference point = $t_1$.
3) The main load is released and the depth of penetration, $t_b$, measured with the preload still applied.

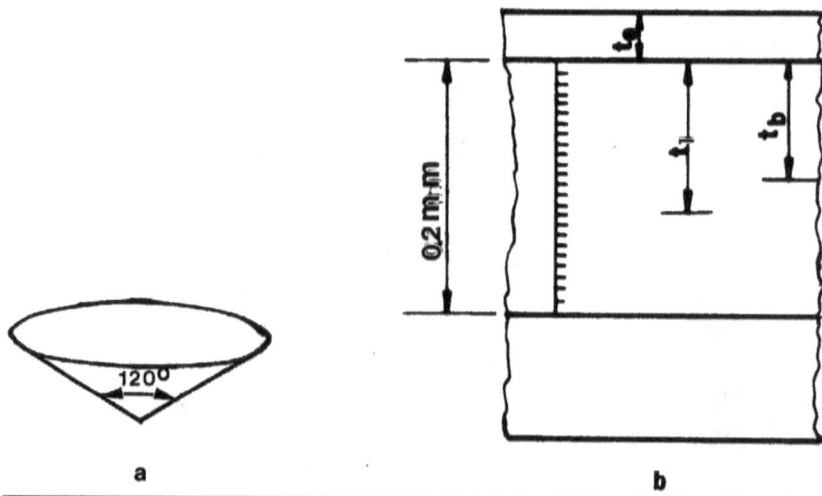

**Fig. 4.8** The Rockwell Hardness test (a) geometry of the indenter (b) the scale

In the case of Rockwell C, the maximum possible penetration depth is 0.2 mm (from the reference surface (see figure 4.8). This is divided into 100 hardness units. Thus the residual penetration, measured in units of 0.002 mm ($0.2\,mm/100$), is:

$$e = t_b/0.002 \text{ hardness units} \tag{4.27}$$
$$\text{Then hardness} = 100 - e \text{ units} \tag{4.28}$$

The hardness determined is then expressed as x HRC, where x is the hardness in arbitrary hardness units and HRC = Rockwell hardness scale C. A similar procedure is followed in the other scales with HRB having the scale divided into 130 units, etc.

In nearly all modern Rockwell hardness testers, a scale is provided which enables the hardness values to be read directly.

Advantages of the Rockwell Scale:
1) The direct reading of the hardness reduces the risk of operator error.
2) The surface does not have to be reflective.
3) May be used to test non-metals e.g., rubbers, bonded materials, etc.
4) The indentations are small and the tested part may be used after the test.
5) High speed test hence very suitable for production line quality control.

Disadvantages:
1) High multiplicity of loads and indenters.
2) Slight sensitivity to heterogeneities due to the small indentation.

## NOTES
1) Hardness values may be <u>approximately</u> converted from one scale to another (tables are available for the approximations) and may be used to <u>estimate</u> the UTS. For steels, UTS (MPa) is about 3.4 HB (kg/mm$^2$).
2) All hardness specimens should be flat and metallurgically bright i.e., able to reflect light (except for Rockwell scale) but curved specimen may be tested and corrections made. Tables for correction factors form part of most standards.
3) Adequate spacing (about 3 x diameter of indentation) is needed between indentations and from the edges of the specimen.

## 4.4 OTHER MECHANICAL PROPERTIES
Other important mechanical properties, which are subject of later study, are:

<u>Toughness</u>: This has been briefly mentioned above and is the ability of a material to absorb energy before fracture. The notch toughness of materials is determined by impact tests as detailed in chapter twelve.

Fatigue strength: This refers to the strength of a material under conditions of continuously varying load. It is considered in more detail in chapter ten.

Creep strength: The response of a material to stresses sustained over long periods of time. It becomes an important consideration in metals at elevated temperatures, and is the subject of chapter eleven.


## PROBLEMS

4.1 (a) Compare and contrast the Vickers and Rockwell scales of hardness testing.

(b) The Vickers hardness of a surgical blade having a thickness of 0.2 mm was determined as 680 HV 10. Check whether the blade thickness was sufficient. (Ans: Not sufficient.)

4.2 A tensile test was performed on a 6.2 mm diameter steel rod with an initial gauge length of 50 mm. The elastic limit was attained at a load of 8.45 kN while the limit of proportionality occurred at a load of 4.53 kN (the corresponding extension being 0.036 mm). If the final gauge length was 83 mm, calculate the modulus of resilience and the logarithmic strain at fracture. (Ans: 188 kJ/m$^3$; 0.51).

4.3 Compare the Vickers and Brinell hardness scales by giving the advantages and disadvantages of each.

4.4 The Vickers hardness of a certain alloy is determined on a 0.5 mm thick plate. A value of 720 HV 30 is read. State (giving reasons) if the thickness of the specimen is sufficient for the test. (Ans: Sufficient.)

4.5 Sketch the stress-strain curves that would be expected when a very ductile material; an elastic-perfectly plastic material; a rigid-perfectly plastic material and an elastic-strain hardening material are tested in a tensile test.

4.6 A Brinell test is performed on a 15 mm thick piece of alloy cast iron using a load of 750 kg and a 5 mm diameter steel ball. An impression diameter of 1.29 mm is measured. Determine with reasons if the value of hardness obtained is valid. If it is not valid, what steps can you take to obtain a valid reading? (Ans: Not valid.)

4.7 During a tensile test on a steel bar, the following were measured: $d_o$ = 6.2 mm, $L_o$ = 50 mm, $d_f$ = 4.8 mm; $L_f$ = 83 mm; P = 8.45 kN and $\Delta L$ = 0.092 mm at the elastic limit; breaking force = 7.31 kN. Calculate:
   (a)    The value of Young's Modulus
   (b)    The modulus of resilience
   (c)    Logarithmic strain at fracture
   (d)    True fracture stress.
(Ans: 152 GPa; 257 kJ/m³; 0.51; 404 MPa.)

4.8 Explain what errors arise when Brinell hardnesses above 450 HB are determined with a steel ball. What steps would you take to avoid the errors?

4.9 The variation of % elongation, δ, with gauge length, $L_o$, of a certain steel follows the following relationship:

$$\delta = \frac{C\sqrt{A}}{L_o} + b$$

where C and b are constants. The following results were obtained on 50 mm x 12.5 mm specimens of the material:

| $L_o$ [mm] | 50 | 100 | 150 | 250 | 300 | 350 |
|---|---|---|---|---|---|---|
| % elong. | 40 | 30.6 | 27.5 | 25.7 | 25 | 24.3 |

Calculate the % elongation for a 50 mm x 22 mm plate of the same steel with a gauge length of 250 mm. (Ans: 26.2 %.)

4.10 (a) Distinguish between the proportionality limit and the elastic limit on the stress-strain curve for metals.

(b) The piston rod of a double acting hydraulic cylinder is 200 mm diameter and 4 m long. The diameter of the piston is 400 mm. It is subjected to 10 MPa water pressure on one side and 3 MPa on the other. On the return stroke, the pressures are interchanged. Determine the maximum compressive stress on the piston rod and the change in length of the rod between the two strokes. E = 200 GPa. (Ans: -31 MPa; 0.98 mm).

4.11 Give two advantages each of the Brinell, Vickers and Rockwell scales of hardness testing.

4.12 During a tensile test on an aluminium specimen, the following results were recorded: $L_o = 70$ mm; $d_o = 14.0$ mm:

| Load [kN] | 0 | 3 | 6 | 9 | 11 | 12.5 | 14 | 15 |
|---|---|---|---|---|---|---|---|---|
| $\Delta L$ [mm] | 0.0 | 0.0197 | 0.0395 | 0.0592 | 0.0724 | 0.096 | 0.125 | 0.185 |

Calculate the modulus of elasticity of the material and the 0.1 % proof stress. If the diameter under the 9.00 kN load was 13.997 mm, determine the material's Poisson's ratio and the diameter under the load of 4.6 kN. (Ans: 69 GPa; 97 MPa; 0.29; 13.998 mm).

4.13 The following results were obtained during a tensile test on a steel specimen with a diameter of 12.8 mm and a gauge length of 50 mm:

| P [kN] | 2.23 | 4.46 | 6.7 | 8.93 | 11.2 | 13.4 | 15.6 |
|---|---|---|---|---|---|---|---|
| $\Delta L$ [mm] | 0.0041 | 0.0081 | 0.013 | 0.018 | 0.023 | 0.027 | 0.031 |

| P [kN] | 17.9 | 20.1 | 22.3 | 24.6 | 26.8 | 28.1 | 31.2 | 33.5 |
|---|---|---|---|---|---|---|---|---|
| $\Delta L$ [mm] | 0.035 | 0.039 | 0.044 | 0.049 | 0.052 | 0.501 | 1.52 | 2.03 |

| P [kN] | 37.9 | 42.9 | 44.6 | 45.1 | 45.5 | 44.9 | 43.1 | 40.6 | 36.2 |
|---|---|---|---|---|---|---|---|---|---|
| $\Delta L$ [mm] | 3.05 | 4.57 | 6.6 | 7.62 | 12.1 | 14.7 | 15.8 | 17.8 | 19.3 |

Lower yield point occurred at 27.7 kN; breaking load 30.4 kN, $L_f = 72.9$ mm; $d_f = 6.76$ mm.

Plot the engineering stress strain curve and hence determine the limit of proportionality; modulus of elasticity; lower yield stress; tensile strength; fracture stress; percent elongation at fracture; percent reduction in area, and the true fracture stress. Plot also the true stress-true strain curve up to maximum load.

4.14 Figure Q4.14 shows the stress strain curve of a steel specimen. The insert is an enlargement of the low strain region of the curve. Determine the following: (a) the modulus of elasticity; (b) the working stress if a factor of safety of 2 against plastic deformation is required; (c) the modulus of resilience; (d) the ultimate tensile strength; (e) the percent elongation at fracture; (f) the permanent deformation if a stress of 800 MPa is applied and then released; (g) the change in length of a specimen, originally 200 mm long, loaded to a stress of 950 MPa; (h) the true stress and true strain at the maximum load. Derive any formula used. (Ans: 206 GPa, 395 MPa, 1.5 MJ/m³; 980 MPa, 15 %, 0.33 %, 8.8 mm, 1 080 MPa, 0.095.)

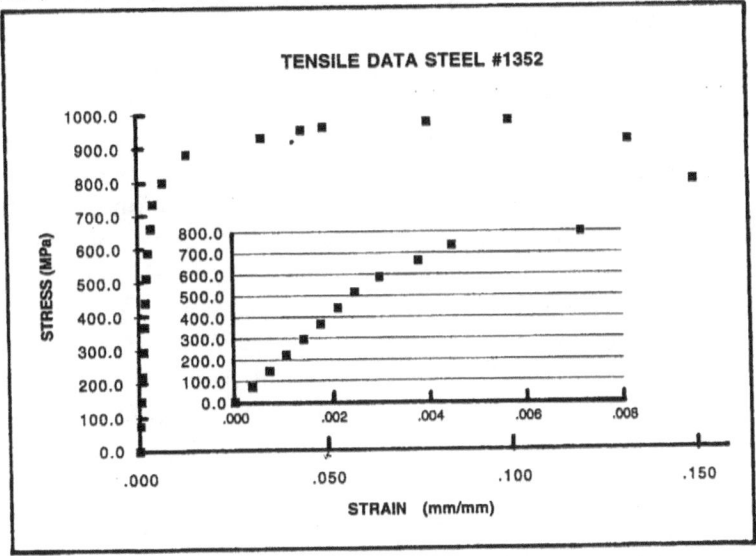

Fig. Q4.14

## Further Reading

Dieter, G. Mechanical Metallurgy, 3 rd. Ed., McGraw-Hill, New York, 1986.

Callister, W. D. Materials Science and Engineering, 3 rd. Ed. John Willey & Sons, New York, 1994.

# CHAPTER FIVE

# EQUILIBRIUM DIAGRAMS AND ALLOY THEORY

## 5.1 ALLOYING SYSTEMS

The properties of metals (with the exception of electrical conductivity, and to some extent the ductility and corrosion resistance) are generally poor in the pure metal. Properties like strength, hardness, toughness, wear resistance, etc., are improved by the addition of other metals or non-metals to the metal in question. A combination of two or more elements in which the major component is a metal into a single material is termed a metallic alloy. Throughout this chapter, the term alloy will be understood to mean metallic alloy. However, it should be pointed out, that the term alloy is sometimes used to refer to mixtures or blends of polymers.

Alloys are formed by melting the alloying elements together. In the liquid state, nearly all metals are miscible in all proportions. The liquid mixture is then cooled and on solidification any of the following may happen:

1. The metals may be completely soluble in the solid state.
2. They may be completely insoluble in solid state.
3. They may be partially soluble in the solid state.
4. An intermetallic compound may form between the metals.

### 5.1.1 Phases

A phase is a portion of matter that is homogeneous in state, crystal structure and chemical composition. According to this definition, a mixture of ice and water constitutes two phases so does a mixture of water and ether. However, a dilute solution of salt in water is a single phase. Matter with more than one phase is called a mixture and has a phase boundary.

### 5.1.2 Solid Solutions

Solid state solutions may form either by substitution or as interstitial solutions. In substitutional solid solutions, atoms of the solute substitute those of the solvent in the latter's crystal lattice. The substitution may be ordered or random. For substitutional solid solution to take place, the atoms of the two elements must be approximately equal in size and have comparable electronic structure. Examples: Zn/Cu (brass), Au/Ag, Cu/Ni.

In interstitial solutions, a comparatively small atom takes the interstice between the atoms of the solvent. (See discussion and diagrams on crystal imperfections, chapter three.)

The bonds in solid solutions are predominantly metallic and the crystal structure remains the same as that of the solvent. Solid solutions are generally soft and ductile, with variable composition and melt over a temperature range.

## 5.1.3 Intermetallic Compounds

Metals form compounds between themselves or with non-metals, which are termed intermetallic compounds. The compounds are hard and brittle with well-defined melting points (solid solutions melt over a range of temperatures). Their crystal structures are complex (with each unit cell having up to 40 atoms or more). The bonds may be ionic, covalent, metallic or mixed. The three types of intermetallic compounds are:

(a) <u>Valency intermetallic compounds:</u> These form between electropositive metals and fairly electronegative metals in groups IV(b), V(b) and VI(b) of the periodic table. Here, the usual rules of valency are followed as in normal compounds and the bonds are either ionic or covalent. Examples: $Mg_3Sb_2$, $Mg_2Sn$, $Mg_3Bi_2$.

(b) <u>Electron intermetallic compounds:</u> These form between metals of comparable electrochemical properties with atoms whose sizes are approximately equal. The rules of valence are not followed but a fixed ratio is maintained. They are usually non-stoichiometric i.e., the same two metals can form several compounds with different proportions. The bonding is essentially metallic. Examples: $CuZn$, $Cu_3Al$, $Cu_5Sn$, $Cu_5Zn_8$, $Cu_9Al_4$, $Ag_5Zn_8$.

(c) <u>Interstitial compounds:</u> Form between metals and non-metals. The small non-metal atoms take the interstice between the metal atoms and are bonded to them (metal atoms) by mixed ionic, covalent and metallic bonds. Examples: $Fe_3C$, $WC$, iron nitrites, etc.

Example 5.1: Table E5.1 gives the atomic radii, crystal structure, the most common valency, and the electronegativities of some elements. State, giving reasons, which element is likely to:

(a) Form a solid solution with nickel showing complete solubility in the solid-state (b) form a substitutional solid solution with nickel (c) form an intermetallic compound with nickel.

Table E5.1

| Element | Atomic radius [nm] | Common valence | Crystal structure | Electro negativity |
|---|---|---|---|---|
| Ni | 0.1246 | 2+ | FCC | 1.8 |
| C | 0.071 | | | |
| H | 0.046 | | | |
| O | 0.060 | | | |
| Ag | 0.1445 | 1+ | FCC | 1.9 |
| Al | 0.1431 | 3+ | FCC | 1.5 |
| Co | 0.1253 | 2+ | HCP | 1.8 |
| Cr | 0.1249 | 3+ | BCC | 1.6 |
| Fe | 0.1241 | 2+ | BCC | 1.8 |
| Pt | 0.1387 | 2+ | FCC | 2.2 |
| Zn | 0.1332 | 2+ | HCP | 1.6 |

Answer:

(a) Complete solubility requires a small difference in atomic radii, same crystal structure, similar eletronegativities and close valency (i.e., a similar electronic configuration). These conditions are best met by platinum.

(b) Ag, Al, Co, Cr, Fe, and Zn will form substitutional solid solution (partial solubility in solid state) with nickel since at least one of the conditions given above is not met.

(c) A small atomic radius is required for a solute to form interstitially. The non-metals, C, H, and O meet this condition.

## 5.2  PHASE DIAGRAMS

Useful engineering alloys are built from solid solutions and intermetallic compounds. In doing this, use is made of equilibrium or phase diagrams which are charts showing the relationships between composition, temperature and structure in an alloy system. They are constituted from cooling curves i.e., the alloying components are melted together in different proportions and then cooled. For each proportion, the temperature at which solidification starts, the temperature at which solidification ends, and the composition of the precipitate, is determined.

The x-axis of the phase diagram represents the composition starting from 100 % metal A to 100 % metal B. The y-axis represents the temperature. Each of the possibilities mentioned in section 5.1 gives a different shape of equilibrium diagram.

## 5.2.1 Complete Solubility in Solid State

Pure metal A will solidify at MP(A) and pure metal B at MP(B) (see figure 5.1). If the melt contains some B, this will raise the solidification temperature of A. The effect increases with increase in the quantity (ratio) of B until we reach MP(B) (assume MP(B) is higher than MP(A)). Furthermore, solidification of mixtures does not take place at a distinct temperature but over a temperature range. We can therefore mark the temperature at which solidification starts and the temperature at which it ends for each composition. The line joining all those points at which solidification starts is termed the liquidus. Above the liquidus, only liquid exists. The line joining the temperatures at which solidification ends is termed the solidus. Below the solidus, we have only solid (the solid solution). In between the solidus and the liquidus, solidification is incomplete hence we have a mixture of solid and liquid.

The equilibrium diagram is then as shown in figure 5.1. This type of reaction can only occur where solid solution is by substitution. Common examples are Cu-Ni; Au-Ag; while less common examples are Bi-Sb; Ag-Pd.

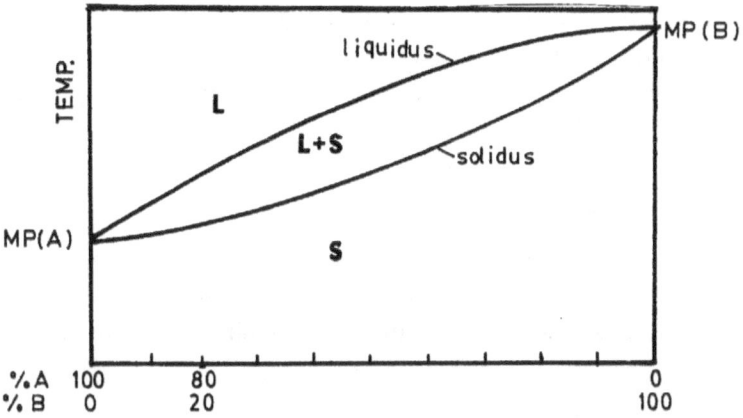

**Fig. 5.1** Phase diagram showing complete solubility in solid state

## 5.2.2 Complete Insolubility in Solid State

Here, the presence of B in A lowers the solidification point of A in a similar manner to the lowering of solidification point of water if it has dissolved salts. Similarly, presence of A in B lowers its (B's) solidification point. When the two sets of points are joined, the liquidii meet at point E (see figure 5.2). On cooling a liquid whose composition is to the left of point E, solidification of A will start at the liquidus. The liquid that is left will be richer in B. This will continue until the liquid left has the composition $C_E$. The same happens with compositions to the right of E but this time, the reaction starts with precipitation of B. At point E, precipitation of both A and B is possible and proceeds as follows:

    1. At some local point, A precipitates.
    2. The surrounding liquid becomes richer in B moving the conditions to the right of E.
    3. B precipitates in the area surrounding the original precipitate of A and conditions shift to the left of E.
    4. A precipitates and the process is repeated.

As a consequence, the resulting precipitate consists of alternate layers of A and B. This constitutes what is called a EUTECTIC. The point E is therefore called the eutectic point. It is the lowest temperature at which the mixture can stay wholly liquid and is important for casting operations. The corresponding composition $C_E$ and the corresponding temperature $T_E$ are termed eutectic composition and eutectic temperature respectively. Since for every composition solidification is complete at point E, the solidus is an isothermal (a constant temperature line) through E. The completed equilibrium diagram is then as shown in figure 5.2.

The reaction at the eutectic point, i.e., the simultaneous precipitation of two solids ($S_1$ and $S_2$) from a liquid, L at the eutectic, point, can be written:

$$L \xrightarrow{\ cooling\ } S_1 + S_2$$
$$L \xleftarrow{\ heating\ } S_1 + S_2 \tag{5.1}$$

This reaction is termed the eutectic reaction, and proceeds to the right on cooling and to the left on heating. Cases of complete insolubility in metals are rare. Examples are calcium-bismuth; silver- silicon; and gold-silicon.

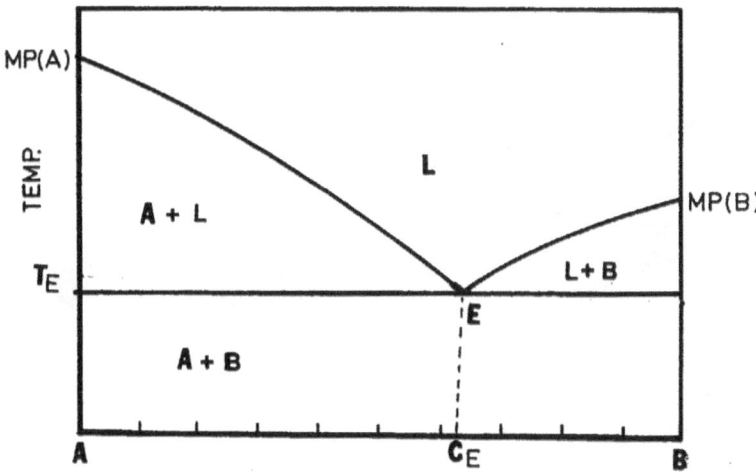

Fig. 5.2 Phase diagram showing complete insolubility in solid state

### 5.2.3 Partial Solubility in Solid State

This case is similar to that of complete insolubility but when A (say) solidifies, it does so with some B dissolved in it. The solid precipitating is termed α (a solution of B in A). Similarly, instead of B precipitating, β = a solution of A in B, solidifies.

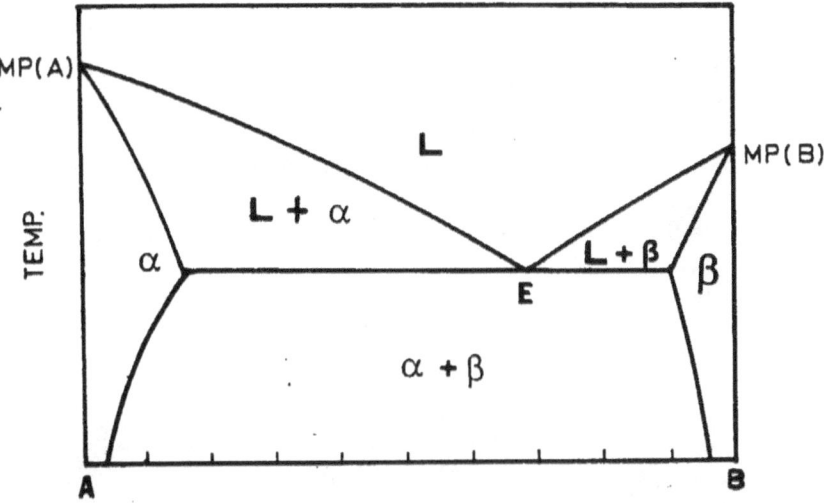

Fig. 5.3 Phase diagram showing partial solubility in solid state

In general (though not always), the solubilities are a maximum at the eutectic temperature. The resulting equilibrium diagram is as shown in

73

figure 5.3. This type of equilibrium diagram is the most common in metals and examples are Ag/Cu; Al/Si; Pb/Sn; Pb/Sb; Bi/Sn; Cd/Pb.

### 5.2.4   Systems with Intermetallic Compound

Let the intermetallic compound be $A_xB_y$. The equilibrium diagram can then be divided into two parts: one between A and $A_xB_y$ and the other between $A_xB_y$ and B. In theory, each of the two parts can be any of the shapes already considered (complete solubility, complete insolubility or partial solubility). In most cases however, neither metal is soluble in the intermetallic compound while there is still some partial solubility of the metals in each other. The most common form of the equilibrium diagram is therefore as shown in figure 5.4.

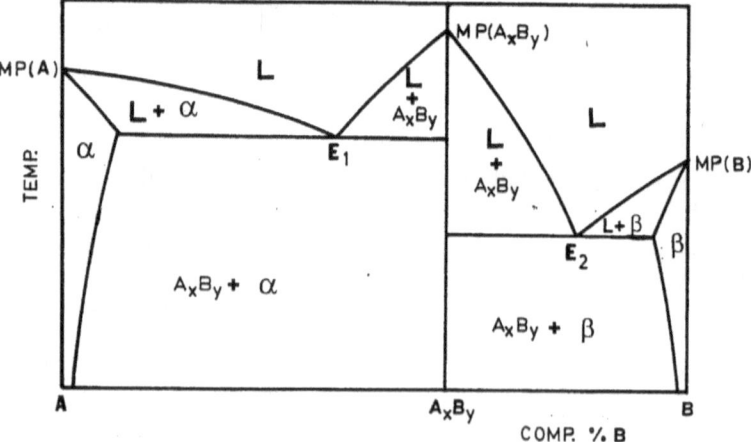

**Fig. 5.4** Phase diagram for a system forming an intermetallic compound

The position of $A_xB_y$ on the equilibrium diagram can be easily found from the atomic weights:

$$wt.\%B = \frac{at.wt.B * y}{at.wt.A * x + at.wt.B * y} \qquad (5.2)$$

Usually, only one portion of the diagram is of technological importance. Examples include Pb/Mg; Cu/Al; Fe/C; and Mg/Sn.

### 5.2.5   Systems with Eutectoid Reaction

In alloying systems where one of the components has a temperature-based polymorphism a change in solubility occurs when the polymorph changes.

The excess solute must then precipitate a new phase resulting in a situation where two solids, $S_2$ and $S_3$ precipitate from one solid, $S_1$ at the change of polymorph. Like the eutectic reaction, the two solids precipitate simultaneously and form alternate layers. This reaction is termed the eutectoid (eutectic like) reaction and may be written:

$$S_1 \xrightarrow{\text{cooling}} S_2 + S_3$$
$$S_1 \xleftarrow{\text{heating}} S_2 + S_3 \qquad (5.3)$$

The best examples are the decomposition of austenite in the Fe-C system and the formation of $\alpha$ and $\gamma_2$ from $\beta$ in the Cu-Al system. Both will be studied further later.

## 5.2.6 Systems with Peritectic Reaction

In some systems, a solid already precipitated, $S_1$ reacts with excess liquid, L to form a new solid, $S_2$. This reaction is termed peritectic reaction and may be written as:

$$S_1 + L \xrightarrow{\text{cooling}} S_2$$
$$S_1 + L \xleftarrow{\text{heating}} S_2 \qquad (5.4)$$

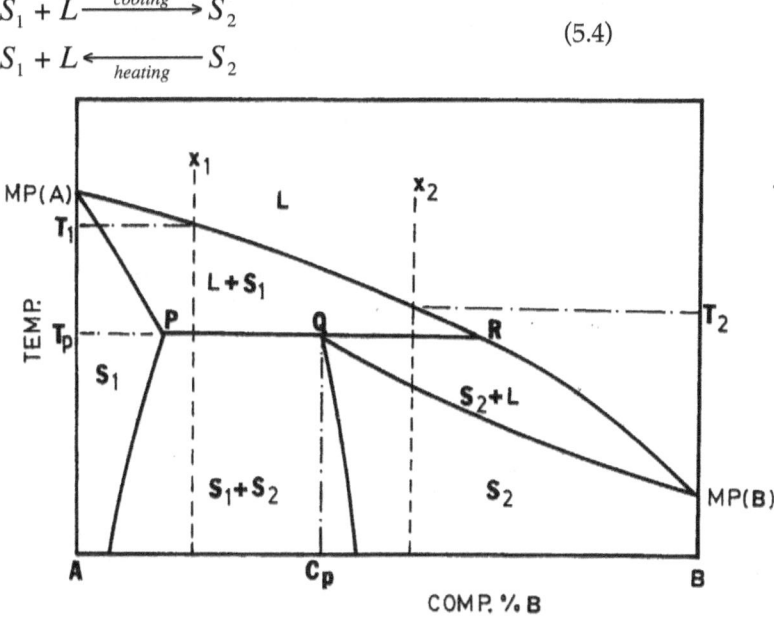

**Fig. 5.5** Phase diagram showing a peritectic reaction

The equilibrium diagram then has the shape shown in figure 5.5 which may be observed to be the inverse of the eutectic reaction (a mirror image of the diagram along PR results in the diagram for the eutectic reaction).

The peritectic reaction can only take place if $S_1$ and liquid are both present together at the peritectic temperature. In figure 5.5, a peritectic reaction is only possible for compositions between P and R. Taking alloy $x_1$, precipitation of A will start at $T_1$ and continue until liquid left has composition R. Here, the peritectic reaction takes place to produce $S_2$ of composition Q. Since there is more $S_1$ than liquid, the liquid is used up leaving some $S_1$. As a result, below $T_p$, we have a mixture of $S_1$ and $S_2$. For alloy of composition $x_2$, $S_1$ will again be deposited until $T_p$. When the peritectic reaction takes place, $S_1$ is used up first. Thus, below $T_p$, we have $S_2$ and liquid. The remaining liquid then precipitates $S_2$ in the normal manner.

## 5.3  COMPOSITION AND QUANTITIES OF PHASES

In two-phase areas, the composition and quantities of the phases are determined as follows:

(i) To determine the chemical composition, an isothermal is drawn through the temperature of interest. The intersection of the isothermal with the phase boundaries gives the compositions. For example the alloy shown in figure 5.6, the composition of solid at $T_1$ is $C_P$ while the composition of liquid is $C_Q$.

(ii) To determine the quantities (or ratios) of the phases, the lever rule is applicable. It states: let the intersection of the isothermal and the line representing the composition (F in figure 5.6) be the fulcrum of a horizontal lever; the lengths of the lever arms to the phase boundaries times the ratios must balance. Thus, in figure 5.6, if $R_S$ = ratio of solid and $R_L$ = ratio of liquid,

$$PF \times R_S \quad = FQ \times R_L \tag{5.5}$$
$$\text{But, } R_L = 1 - R_S \tag{5.6}$$
$$\Rightarrow \quad PF \times R_S = FQ\,(1 - R_S)$$
$$R_S\,(PF + FQ) = FQ$$
$$\text{Or, } R_S = {}^{FQ}/_{PQ} \text{ (since } PF + FQ = PQ) \tag{5.7}$$
$$\text{Similarly, } R_L = {}^{PF}/_{PQ} \tag{5.8}$$

These can be easily remembered as "opposite"/"total".

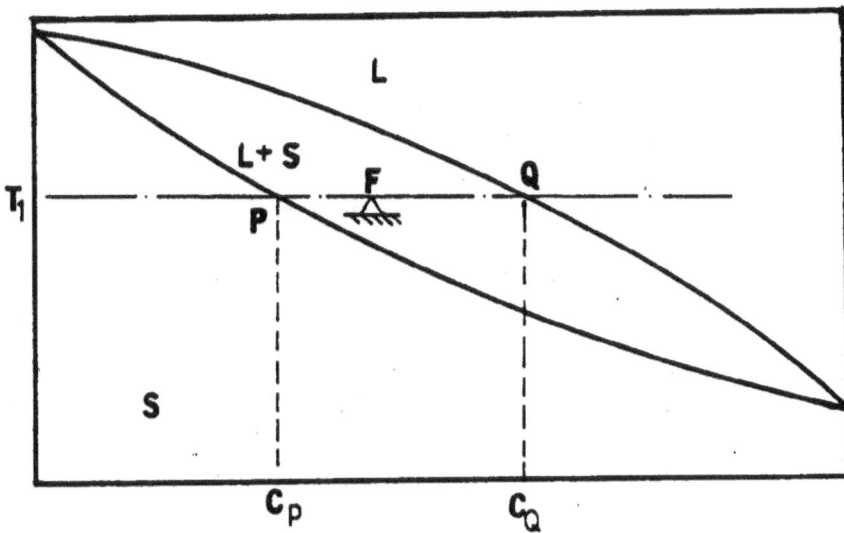

**Fig. 5.6** Composition and quantities of phases

Example 5.2: Draw the silver-copper equilibrium diagram (between 1100 °C and 400 °C) given the following data:

| | |
|---|---|
| Melting point of silver | 962 °C |
| Melting point of copper | 1084 °C |
| Eutectic temperature | 780 °C |
| Eutectic composition | 28.1 % Cu |
| Solubility of Cu in Ag at eutectic temperature | 8.8 % |
| Solubility of Ag in Cu at eutectic temperature | 8.0 % |
| Solubility of Cu in Ag at 400 °C | 2.0 % |
| Solubility of Ag in Cu at 400 °C | 0.0 % |

You may use straight lines for phase boundaries. For 360 g of an alloy containing 80 % copper, determine: (a) the temperatures at which melting starts and ends (b) the phases present, their chemical composition and their quantities at 950 °C.

Answer:

It is clear from the information given that this is a case of partial solubility in the solid state. If $\alpha$ = solution of Ag in Cu and $\beta$ = solution of Cu in Ag, the phase diagram is as shown in figure E5.2.

(i) For the alloy in question, melting starts at 780 °C and ends at 1000 °C.

(ii) At 950 °C, phases present are $\alpha$ and liquid, L. Chemical compositions are: $\alpha$: 96.5 % Cu, 3.5 % Ag. Liquid: 68.5 % Cu, 31.5 % Ag.

$$\alpha = \frac{23}{56} x360g = 148g$$

L = 360 - 148 = 212 g

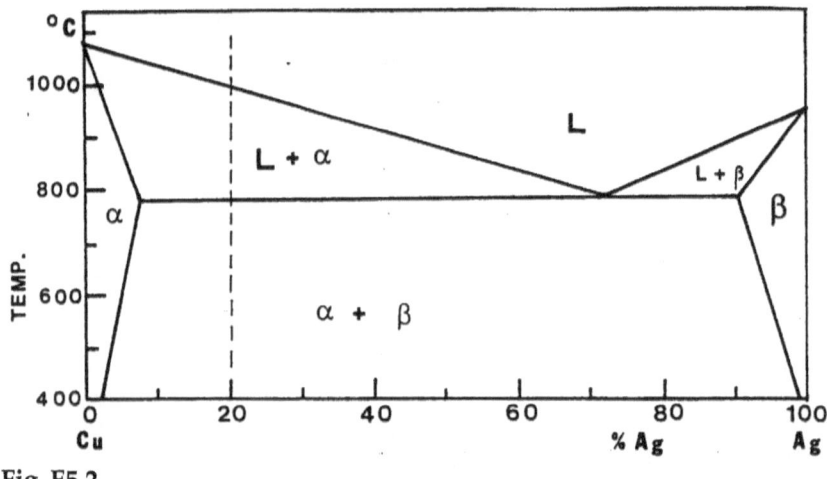

**Fig. E5.2**

Example 5.3: Two hypothetical metals A and B form an intermetallic compound $A_2B_3$. Given the following data, draw and fully label the phase equilibrium diagram between the two metals.

| | |
|---|---|
| Atomic weight of A | 62.5 amu |
| Atomic weight of B | 30.3 amu |
| Melting point of A | 730 °C |
| Melting point of B | 580 °C |
| Melting point of $A_2B_3$ | 1000 °C |
| Eutectic temperature, A-$A_2B_3$ portion | 600 °C |
| Eutectic temperature, $A_2B_3$-B portion | 480 °C |
| Eutectic composition, A-$A_2B_3$ portion | 30 % B |
| Eutectic composition, $A_2B_3$-B portion | 70 % B |
| Maximum solubility of B in A | 0 % |
| Maximum solubility of A in B | 8 % |
| Solubility of A in B at 0 °C | 3 % |

Use straight lines for all phase boundaries. For 100 g of an alloy containing 95 % B, determine: (a) the temperatures at which melting starts and ends (b) the phases present, their chemical composition and their quantities at 520 °C.

Answer:

This is a case of a system forming an intermetallic compound. The position of $A_2B_3$ on the x-axis is first determined from equation 5.2

$$wt.\%B = \frac{30.3x3}{2x62.5 + 3x30.3} = 42\%$$

Define $\alpha$ = solution of B in A, and $\beta$ = solution of A in B. The phase equilibrium diagram is shown in figure E5.3.

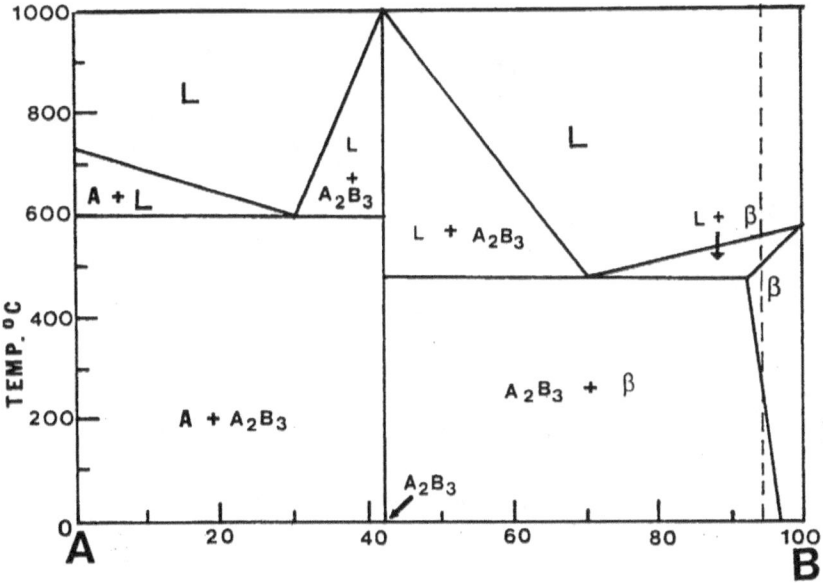

Fig. E5.3

(a) For the alloy in question, melting starts at 510 °C and ends at 545 °C

(b) Phases present at 520 °C are L and $\beta$
    Compositions:  L - 15 % A, 85 % B
                        $\beta$ - 4 % A, 96 % B
    Quantities:      L - 9.1 g
                     $\beta$ - 100 - 9.1 = 90.9 g

## 5.4  COOLING CURVES

Phase diagrams are constructed from cooling curves. These are simply plots showing the variation of temperature with time as a specific mixture is cooled. Whereas a pure metal solidifies at a fixed temperature (which is indicated by a horizontal portion of a cooling curve), an alloy solidifies over a temperature range (between the liquidus temperature and the solidus temperature). However, a change in slope of the cooling curve will occur at these respective temperatures. Solidification of alloys having eutectic and eutectoid compositions will also take place at constant temperature. Figure 5.7 shows the general shape of the cooling curves expected from a system showing complete insolubility in the solid state.

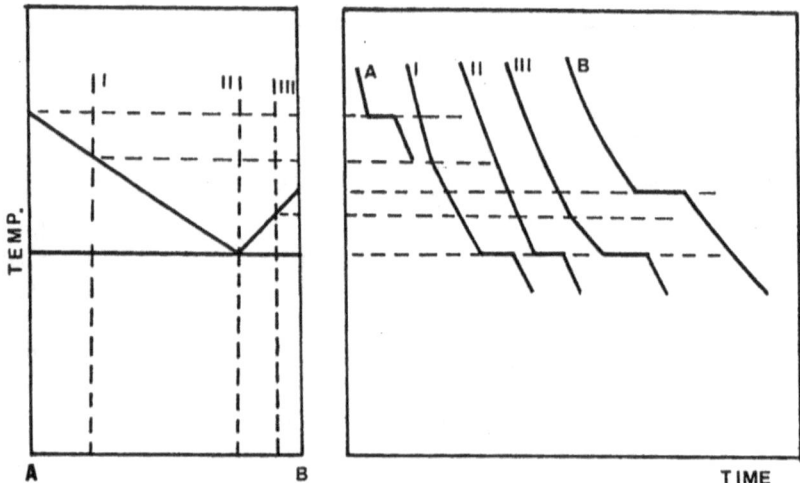

**Fig. 5.7** An example of cooling curves and the corresponding phase diagram

## NOTES

### 1. DENDRITES

Solidification from the melt occurs by nucleation and growth. Tiny nuclei form at various parts of the melt and then grow at the expense of the liquid surrounding them. The growth may be more rapid along certain crystallographic directions resulting in branched crystals called dendrites. When dendrites of different orientation touch, growth stops. The rest of the melt then solidifies and the contact surfaces become the grain boundaries.

2. CORING.

From what was said in section 5.3 on chemical composition, it is clear that the precipitating solid has a different composition as the temperature decreases. For example, in figure 5.1, the first solid to precipitate will have a higher proportion of B than the last. The resulting grain or crystal will thus have varying composition from the centre (more B) to the edges (less B) and is said to be cored. Coring can only be avoided (or reduced) if the cooling rate is extremely low to allow diffusion to take place. Cored casting have inferior mechanical properties compared to homogeneous ones and hence cored casting are sometimes re--heated to just below the solidus temperature and kept at this temperature to allow diffusion to take place. This procedure is termed homogenization.

## PROBLEMS

5.1 Sketch and Label the equilibrium diagram between two metals which show complete solubility in the solid state. What factors would favour this type of behaviour?

5.2 Sketch and label the equilibrium diagram between two metals A and B given that A and B form an intermetallic compound in which both metals are insoluble. A and B are also partially soluble in each other. Identify all phases in your sketch.

5.3 Sketch and label the equilibrium diagram, between two metals A and B in the hypothetical case where A and B form non-stoichiometric (i.e., with variable composition) intermetallic phase. A and B are also partially soluble in each other. Define all phases in your sketch.

5.4 What are the differences between intermetallic compounds and solid solutions with regard to composition, type of bonds, crystal structure and mechanical properties?

5.5 Sketch schematically and fully label the phase diagram of a binary alloy system showing a peritectic reaction. Give one common example of an alloy system showing this type of diagram.

5.6 Referring to the silver-copper equilibrium diagram given in example 5.2, draw the cooling curves expected from alloys with the following compositions: 5 % silver; 8.0 % silver; 40 % silver; 71.9 % silver; and 100 % silver. Indicate the temperatures at which the changes in slope occur.

5.7 What are the differences between valency and electron intermetallic compounds as regards their bonding, valence rules, elements forming the compound and composition?

5.8 Two metals A and B are partially soluble in the solid state. The maximum solubilities by weight are 5 % B and 25 % A. The solubilities are 2 % and 5 % respectively at 0°C. The metals form an intermetallic compound $A_2B$ melting at 750 °C. The respective melting points of A and B are 700 °C and 500 °C while the atomic weights are 45 and 75 respectively. Eutectics are formed at 22 % and 60 % by weight of B at temperatures of 450 °C and 320 °C respectively.

Construct and fully label the equilibrium diagram using straight lines for all phase boundaries.

For 220 g of an alloy containing 45 % A, determine the temperature at which melting starts and ends on heating and the composition and quantities of phases at 200 °C. [Ans: 320 °C; 460 °C; β (83 % B); $A_2B$ (45 % B), 162 g; 58 g].

5.9 A binary alloy is made between two metals A and B. Given the following data, draw and fully label the equilibrium diagram:

| | |
|---|---|
| Melting point A | 700 °C |
| Melting point B | 1000 °C |
| Maximum solubility of A in B | 15 % |
| Maximum solubility of B in A | 10 % |
| Eutectic temperature | 500 °C |
| Eutectic composition | 40 %B |
| Solubility of A in B at 0 °C | 10 % |
| Solubility of B in A at 0 °C | 5 % |

For 500 g of an alloy containing 30 % B; determine the chemical composition and the quantities of the phases at 550 °C. [Ans: L (30 % B); 500 g.]

5.10 (a) Explain, with illustrations where possible, the meanings of: eutectic alloy; peritectic reaction; lever rule; phase.

(b) There are 500 g of an alloy consisting of two components A and B. The metals, A and B are completely soluble in the liquid state but only partially soluble in the solid state. Given the following data, draw and fully label the equilibrium diagram (you may use straight lines for simplicity):

| | |
|---|---|
| Melting point of A | 650 °C |
| Melting point of B | 550 °C |
| Eutectic composition | 40 % A |
| Eutectic temperature | 350 °C |
| Maximum solubility of B in A (at 350 °C) | 15 % |
| Maximum solubility of A in B (at 350 °C) | 10 % |

Solubility at 0 °C = 0 % in both cases.

For an alloy consisting of 75 % A, calculate the quantities and the chemical compositions of the phases present at 400 °C and 350 °C. [Ans: α (143 g, 12.5 % B); L (357 g; 50 % B); β (90 % B); L (60 % B). Quantities are indeterminate at 350 ºC.]

5.11 (a) Explain, with examples where possible, the meanings of the following terms: eutectoid alloy, peritectic reaction, substitutional solution, valency intermetallic compound.

(b) Figure Q5.11 shows a phase diagram between two metals, A and B, soluble in the liquid state but only partially soluble in the solid state. For alloys I, II, and III, explain the crystal structures that you would expect at temperatures $T_1$, $T_2$, $T_3$, $T_4$, and $T_5$.

5.12 Draw the equilibrium diagram between lead (Pb) and tin (Sn) given the following data:

| | |
|---|---|
| Melting point of Pb | 327 ºC |
| Melting point of Sn | 232 ºC |
| Eutectic temperature | 183 ºC |
| Eutectic composition | 61.9 % Sn |
| Solubility of Sn in Pb at 150 ºC | 10 % |
| Solubility of Sn in Pb at 50 ºC | 1.5 % |
| Maximum solubility of Sn in Pb | 19.2 % |
| Maximun solubility of Pb in Sn | 2.5 % |
| Solubility of Pb in Sn at 132 ºC | 0 % |

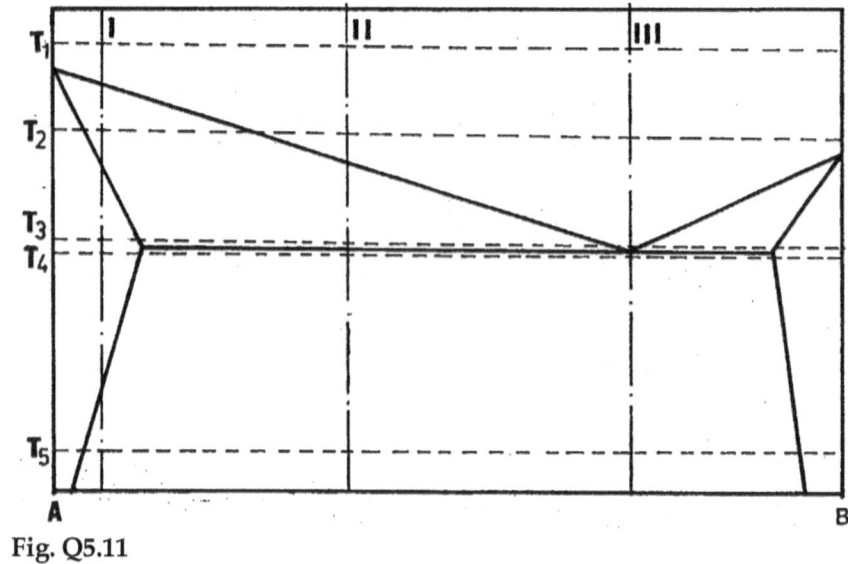

**Fig. Q5.11**

You may use straight lines for all phase boundaries where no alternative data is given.

For 700 g of an alloy containing 83 % Sn, calculate the quantities of phases present and their chemical compositions at 200 °C. [Ans: L (457 g, 75 % Sn); β (98 % Sn, 243 g)].

5.13 Draw and label the peritectic phase diagram between silver and platinum (between 1900 °C and 800 °C) given the following data: (use straight lines for phase boundaries).

| | |
|---|---|
| Melting point of silver | 960 °C |
| Melting point of platinum | 1770 °C |
| Peritectic temperature | 1189 °C |
| Peritectic composition | 54 % Pt |
| Composition of liquid at peritectic temperature | 31 % Pt. |
| Composition of solid at peritectic temperature | 86 % Pt. |
| Compositions of the solid solutions at 900 °C (respectively) | 57 % Pt & 88 % Pt. |

From the diagram you have drawn, determine the following:
(i) For an alloy containing 28 % Pt., the temperature at which melting starts and ends on heating.

(ii) For an alloy containing 36 % Ag, the phases present, their chemical composition and their ratios at 1400 °C. [Ans: 1090 °C; 1175 °C; L (43 % Ag); S (9 % Ag); 0.78; 0.22].

5.14 Draw and fully label the equilibrium diagram between lead and magnesium given the following data:

| | |
|---|---|
| Intermetallic compound | PbMg$_2$ |
| Atomic weight of Pb | 207 |
| Atomic weight of Mg | 24 |
| Melting point of magnesium | 649 °C |
| Melting point of lead | 328 °C |
| Melting point of PbMg$_2$ | 550 °C |
| Eutectic temperature of Pb/PbMg$_2$ portion | 250 °C |
| Maximum solubility of Mg in Pb | 1 % |
| Eutectic composition of PbMg$_2$/Mg portion | 32 % Mg |
| Eutectic composition of Pb/PbMg$_2$ portion | 2.5 % Mg |
| Eutectic temperature of PbMg$_2$/ Mg portion | 468 °C |
| Maximum solubility of Pb in Mg | 26 % |

Both solubilities (Mg in Pb and Pb in Mg) are approximately 0 % at 0 °C and both metals are insoluble in PbMg$_2$.

For 300 g of an alloy containing 35 % Mg, determine the quantities of the phases present at 370 °C. Use straight lines for all phase boundaries. [Ans: β: 77.5 g; PbMg$_2$: 222.5 g].

5.15 Magnesium and tin form a valency intermetallic compound Mg$_2$Sn. Given the following data, draw and fully label the equilibrium diagram:

| | |
|---|---|
| Atomic weight of magnesium | 24.3 |
| Melting point of magnesium | 649 °C |
| Melting point of tin | 232 °C |
| Atomic weight of tin | 118.69 |
| Maximum solubility of Sn in Mg | 15 % |
| Solubility of Sn in Mg at 400°C | 4.4 % |
| Solubility of Sn in Mg at 100°C | 0 % |
| Eutectic temperature of Mg rich portion | 561 °C |
| Eutectic composition of Mg rich portion | 36.4 % Sn |
| Melting point of Mg$_2$Sn | 772 °C |

| Maximum solubility of Mg in Sn | 0 % |
|---|---|
| Eutectic temperature of Sn rich portion | 204 °C |
| Eutectic composition of Sn rich portion | 98 % Sn |

Neither metal is soluble in $Mg_2Sn$. You may use straight lines for phase boundaries where no alternative data is given.

For 500 g of an alloy containing 22 % Sn, which is at a temperature just below 561°C, calculate:

(i) The quantity of pro-eutectic $\alpha$ (where $\alpha$ is the solution of Sn in Mg). (Ans: 336.4 g.)

(ii) The quantity of eutectic $\alpha$ (Ans: 101 g.)

(iii) The quantity of $Mg_2Sn$. (Ans: 62.6 g.)

## Further Reading:

Higgins, R.A. Engineering Metallurgy Part I, Holder & Stoughton, London, 1980.

Anderson, J.C. et al. Materials Science, 3rd ed., ELBS/Van Nostrand Reinhold, UK, 1985.

Van Vlack, L. H. Materials for Engineering, Addison-Wesley, Reading, 1982.

Schaffer, J. P. et al. The Science and Design of Engineering Materials, Irwin, Chicago, 1995.

# CHAPTER SIX

# THE IRON-CARBON EQUILIBRIUM DIAGRAM AND THE FERROUS ALLOYS

## 6.1 INTRODUCTION

The Iron-carbon diagram is the basis of two very important groups of alloys: steels and cast irons (also termed ferrous alloys). Their importance derives from the following:

(i) The relative abundance of iron, which forms about 5 % of the earth's crust.

(ii) The ease with which the iron ore can be reduced to the metal. The reduction can be done using carbon (which is cheap and is also the alloying element) as the reducing agent.

(iii) Their wide range of properties that make them very versatile.

For this reason, ferrous alloys are relatively cheap and are the first choice as structural material unless other considerations (weight, corrosion resistance, etc.,) dictate otherwise. Their main disadvantage is their high susceptibility to atmospheric corrosion.

## 6.2 ALLOTROPIC FORMS OF IRON

Iron exhibit temperature based polymorphism (see chapter two): up to 912 °C, it has a BCC crystal structure and this polymorph is termed α-iron. Between 912 °C and 1400 °C, the structure changes to FCC. The polymorph is termed γ-iron. The last polymorph, δ-iron, which also has a BCC structure, exists between 1400 °C and 1535 °C, (the melting point of iron). Therefore, on heating, the volume change of iron with temperature will have the shape shown in figure 6.1. There is normal thermal expansion up to 912 °C when a sudden drop occurs. This results due to the closer packing of the FCC structure. A sudden rise in volume takes place at 1400 °C when the structure reverts to BCC. This is followed by normal thermal expansion up to melting at 1535 °C.

Carbon dissolves interstitially in all the above polymorphs of iron. It has a maximum solubility of 0.02 % in α-Fe at 723 °C. The resulting solid solution is given the special name ferrite and designated by α. The solubility in γ-Fe reaches a maximum of 2.06 % at 1146 °C, the resulting solid solution being given the name austenite and is designated by γ.

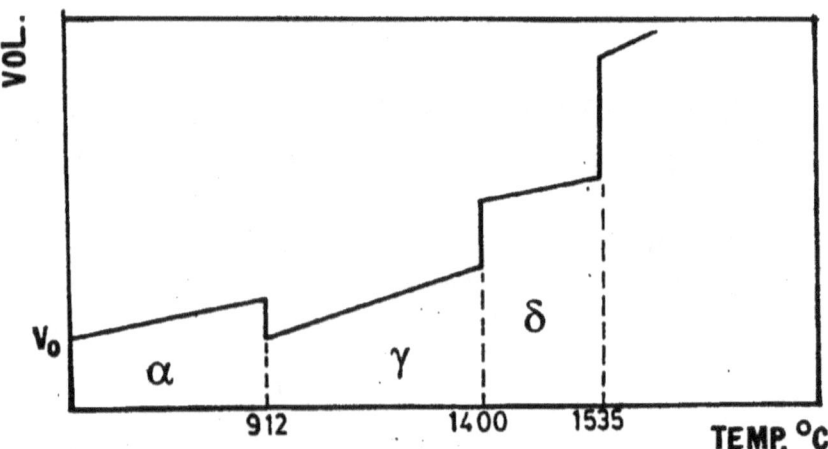

**Fig. 6.1** Schematic diagram showing variation of volume of iron with temperature

The higher solubility of carbon in γ-Fe (despite its higher atomic packing factor compared to α-Fe) results from the fact that the interstices in FCC are fewer but each one is bigger than those in the BCC (which are more but smaller).

Finally, δ dissolves up to 0.1 % carbon at 1492 °C. This solution has no technological importance and hence has no special name. It is designated by δ.

In addition to the above solid solutions, iron and carbon form an intermetallic compound $Fe_3C$, which is termed cementite or carbide. From equation 5.2, the weight percentage of carbon in $Fe_3C$ is 6.67 %.

## 6.3  EUTECTIC, EUTECTOID AND PERITECTIC DECOMPOSITION IN THE FE-C SYSTEM

Like most systems with an intermetallic compound, only the part of the Fe-C diagram between Fe and $Fe_3C$ has technological importance. In the Fe-$Fe_3C$ system, the peritectic reaction takes place at 1492 °C (0.3 % C) when δ reacts with excess liquid to form γ. The eutectic reaction takes place at 1146 °C and 4.3 % C when the liquid precipitates alternate layers of austenite and cementite. The cementite that results from the eutectic reaction is termed primary cementite and the eutectic is termed ledeburite. The eutectoid reaction takes place at 723 °C, 0.8 % C when iron changes polymorph from γ-Fe (FCC) to α-Fe (BCC). Since the solubility of carbon in γ-Fe is much higher than in α-Fe, the excess carbon diffuses out of solution to form Fe₃C. Therefore, we have γ precipitating alternate layers of α and cementite. The product of the eutectoid reaction is termed pearlite.

With the information given, the iron-carbon (or more correctly the Iron-Fe₃C) diagram looks as shown in figure 6.2. A magnified part of the low carbon part (up to 2.1 % C) is shown in figure 6.3. This is the part relevant to steels. Due to the importance of ferrous alloys all mechanical engineers must be well conversant with the Fe-C diagram.

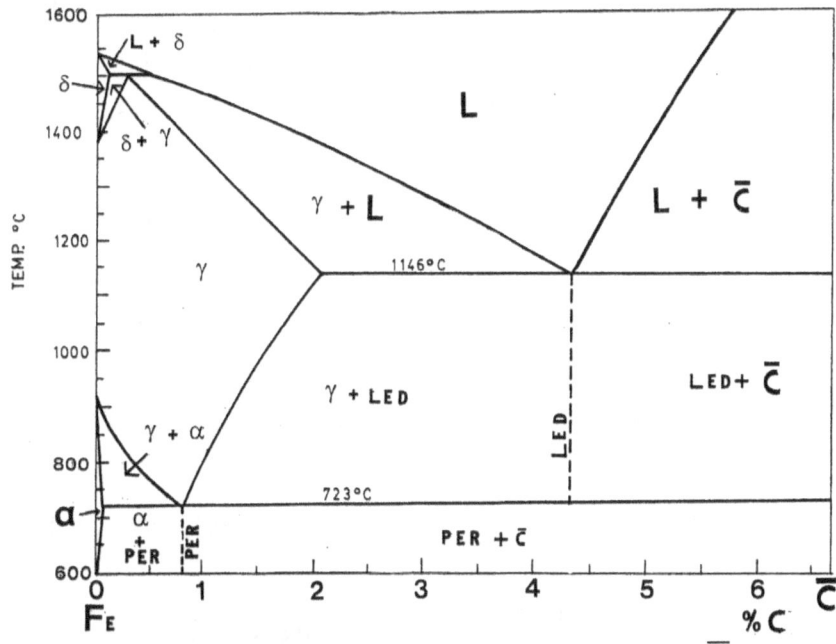

**Fig. 6.2.** The Fe-Fe₃C equilibrium diagram. Led = Ledeburite, $\overline{C}$ = Cementite, Per = Pearlite

## 6.4 STABLE AND METASTABLE FE-C SYSTEMS

The Fe-Fe₃C system described in section 6.3 is termed the metastable system, since Fe₃C is a metastable phase, which decomposes under certain conditions to ferrite and carbon in the form of graphite. These conditions are slow cooling rate, high temperature and presence of other elements especially silicon and/or nickel. Even though graphite (carbon) is the stable phase in equilibrium with iron, the decomposition of cementite is negligibly slow at room temperature. The Fe-graphite system is referred to as the stable system. The stable Fe-C diagram looks similar to the metastable diagram except for small shifts in the position of the invariant points and the isotherms. It is shown in figure 6.4.

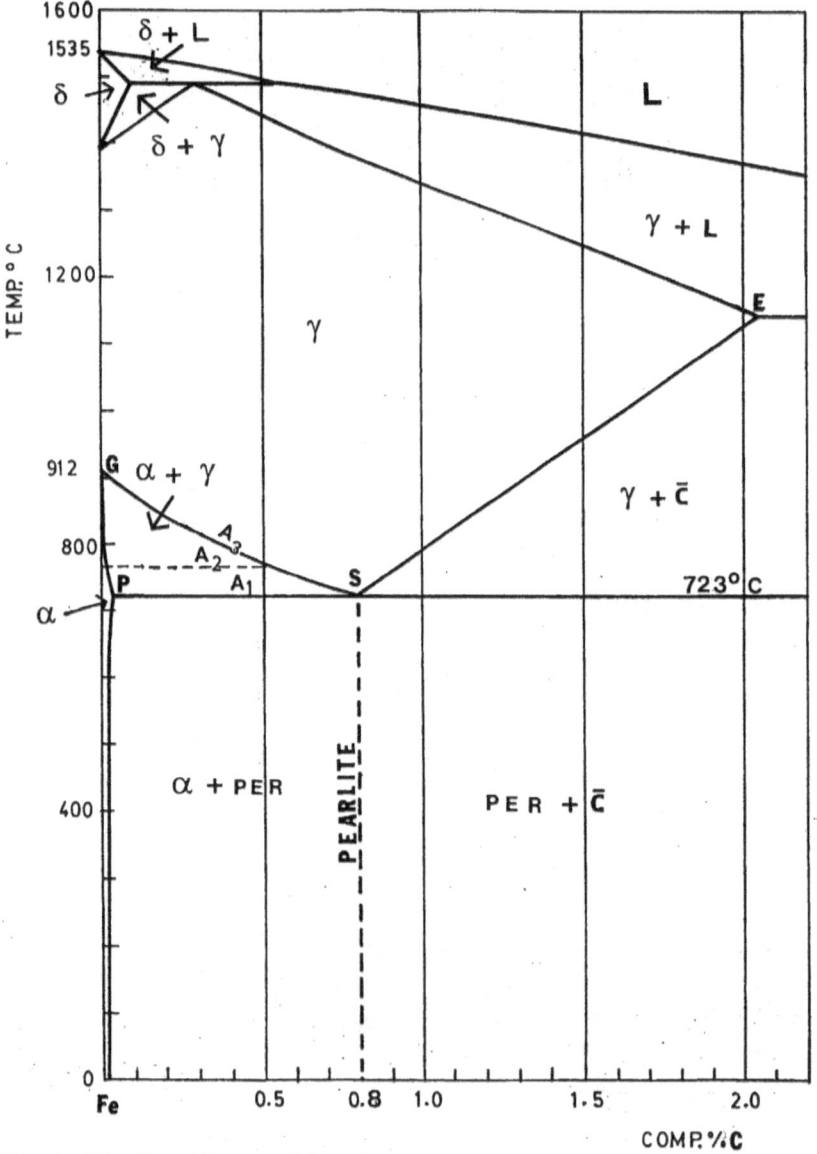

**Fig. 6.3** The "steel" part of the Iron-cementite diagram

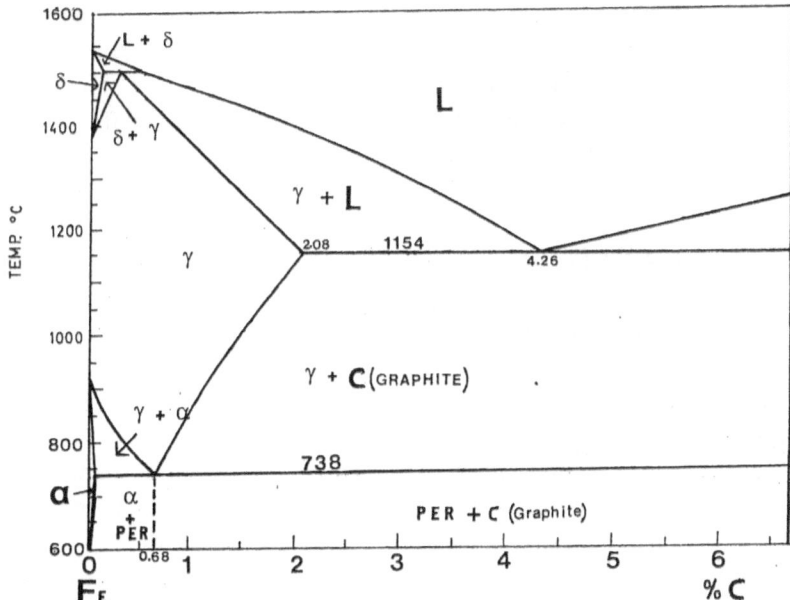

**Fig. 6.4** The Fe-graphite phase diagram

## 6.5  PLAIN CARBON STEELS
### 6.5.1  Terminology

The products of the eutectoid reaction are termed steels. They have less than 2.06 %C and have the distinct characteristic that they can be heated to a single solid phase γ. This makes them suitable for hot working processes like forging, hot rolling, etc. As will be detailed in chapter nine, the properties of steels can be varied over a wide range by controlling the rate of the decomposition of austenite to ferrite and cementite, and by adding other alloying elements in addition to carbon. This accounts for the very widespread use of steel.

Steels that contain only carbon as alloying element (with up to 1 % Mn) is termed a plain carbon steel. If any other alloying elements are added, it is termed alloy steel. More about alloy steels will be considered in chapter nine. The following terminology is commonly used when dealing with plain carbon steels:

(i)    Ingot iron has carbon content less than 0.05 %.

(ii)   Dead soft mild steels have carbon content between 0.05 % and 0.15 %.

(iii)  Mild steels have carbon content between 0.15 % and 0.3 %. These have adequate plasticity, good machinability and weldability.

Mild steel is therefore the most commonly used structural material and is used, among other things, for making structural shapes: I-beams, channels and angles.

(iv) Medium carbon steels have carbon content between 0.3 % and 0.5 %.

(v) High carbon steels have carbon content higher than 0.5 %.

## 6.5.2 Hypo and Hyper-eutectoid Steels

As stated above, steels are the product of the eutectoid reaction in the Fe-C system. The eutectoid reaction gives simultaneous precipitation of α and cementite from γ. The resulting product, which consists of alternate layers of α and cementite, is termed pearlite (from its appearance of alternate black and white streaks when viewed under the microscope). Steels with 0.8 % C consists entirely of pearlite when cooled under equilibrium conditions and is termed eutectoid steel. Steel with carbon content below 0.8 % is termed hypo-eutectoid steel. If cooled under equilibrium conditions, the microstructure of hypo-eutectoid steels shows a mixture of ferrite (which precipitates before the eutectoid reaction and is termed pro-eutectoid ferrite) and pearlite. The ferrite and cementite in pearlite are termed eutectoid ferrite and eutectoid cementite, respectively. Steel with carbon content in excess of 0.8 % C is termed hyper-eutectoid steel. When viewed under a microscope, it shows pro-eutectoid or secondary cementite (usually precipitated at the grain boundaries of austenite so called prior-austenite grain boundaries) and pearlite. Figure 6.5 shows schematically the grain structure of the various steels. If the cooling rate is rather fast, pro-eutectoid (secondary) cementite may precipitate inside the grains along certain preferred crystal plains. The resulting microstructure shows a needle like network of cementite inside a pearlite matrix, which is termed a Widmanstatten structure.

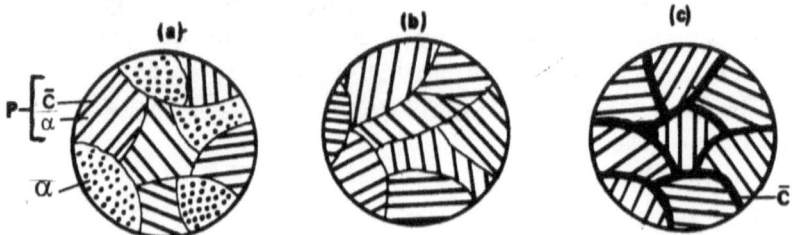

**Fig. 6.5** Schematic structures: (a) hypo-eutectoid (b) eutectoid (c) hyper-eutectoid steels.

### 6.5.3 Quantities of Phases in the Fe-C System

The ratios of the phases in the Fe-C system can be easily calculated using the lever rule. For hypo-eutectoid steels with x % carbon just below the eutectoid temperature:

Ratio of pro-eutectoid ferrite:

$$\frac{0.8 - x}{0.8 - 0.02} \qquad (6.1)$$

Ratio of pearlite = ratio of γ just above eutectoid temperature (figure 6.3). This is given by:

$$\frac{x - 0.02}{0.8 - 0.02} \qquad (6.2)$$

Ratio of eutectoid ferrite:

$$\frac{6.67 - 0.8}{6.67 - 0.02} = 0.88 \qquad (6.3)$$

Eutectoid cementite:
$$= 0.12 \text{ of quantity of Pearlite} \qquad (6.4)$$
The ratio of total ferrite (pro-eutectoid + eutectoid):

$$\frac{6.67 - x}{6.67 - 0.02} \qquad (6.5)$$

And that of cementite:

$$\frac{x - 0.02}{6.67 - 0.02} \qquad (6.6)$$

Similarly for a hyper-eutectoid steel containing x % C

Ratio of pro-eutectoid cementite:

$$\frac{x - 0.8}{6.67 - 0.8} \qquad (6.7)$$

Ratio of pearlite:

$$\frac{6.67 - x}{6.67 - 0.8} \qquad (6.8)$$

The ratios of eutectoid ferrite and cementite can be calculated from the same equations as for hypo-eutectoid steel.

Example 6.1: Calculate the quantities of pro-eutectoid ferrite, pearlite, eutectoid ferrite and cementite in 1 kg of a 0.3 % C steel at 722 °C, assuming equilibrium conditions.

From equation 6.1 setting x = 0.3:

$$wt.\alpha = \frac{0.8 - 0.3}{0.8 - 0.02} x1000g = 641g$$

Quantity of pearlite:    1000 g - 641 g = 359 g

Quantity of pro-eutectoid a: 0.88 x quantity of pearlite = 0.88 x 359 g = 316 g.

Quantity of cementite: 0.12 x quantity of pearlite = 0.12 x 359 g = 43 g

As the temperature changes, the ratios also change and hence for any steel, a phase transformation diagram can be plotted. Such a diagram is shown in figure 6.6 for 0.6% C steel. Phase transformation diagrams are easily constructed from the phase diagram using equations 6.1 to 6.8 as appropriate.

Decrease in the solubility of carbon in ferrite, as the temperature is decreased lead to precipitation of ternary cementite that segregate to the grain boundaries or form plate-like needles inside the ferrite matrix.

Quantities of phases in the stable system and for products of the eutectic reaction can be determined easily in a manner similar to the procedure used for steels. If the reader has followed the logic used in deriving equations 6.1 to 6.8 this should be straightforward.

## 6.5.4   Properties of Annealed Plain-Carbon Steels
The mechanical properties of plain carbon steels are affected by:
   (i) The size and number of phases present;
   (ii) The ratio of the phases (which is a function of the carbon content);
   (iii) Shape and distribution of the minor phase.

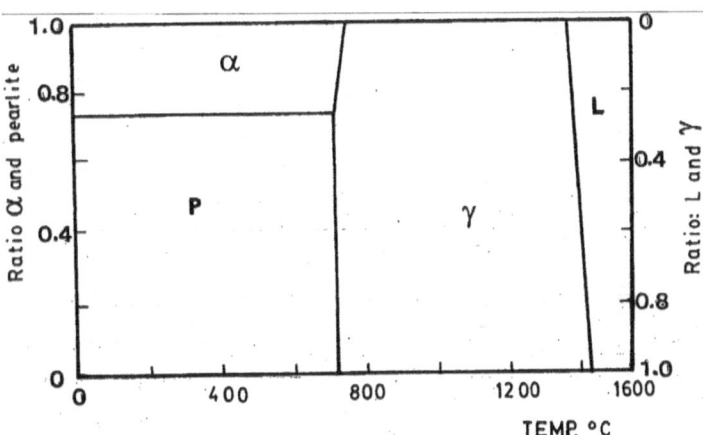

**Fig. 6.6** Phase transformation diagram for 0.6 % C steel

Since ferrite is a solid solution it is soft and ductile. Pearlite has moderate strength, which depends on the size of the lamellae. The size of the lamellae in turn depends on the cooling rate: The slower the cooling rate, the coarser the lamellae and the lower the hardness (approximately 170 HB). Finer lamellae can have hardness up to 340 HB since slip is more difficult in the finer structure.

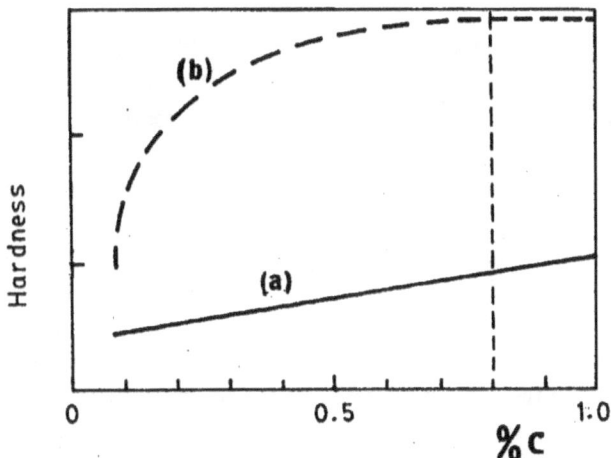

**Fig. 6.7** The variation of hardness with carbon content in plain carbon steels (a) normalised (b) quenched

When other factors are kept constant therefore, the strength (hardness) of steel will vary with carbon content as shown in figure 6.7. As the carbon

95

content increases, the ratio of pearlite increases. Maximum hardness is achieved at 100 % pearlite (0.8 % C).

The shape and distribution of the minor phase also affects the properties. If the minor phase (cementite in this case) is distributed as fine particles inside the ductile ferrite, the toughness is increased. If the particles are larger or if they segregate to the grain boundaries, both the hardness and toughness are reduced.

## 6.6 THE CAST IRONS

The products of the eutectic reaction (1146°C, 4.3 %C) are termed cast irons. They contain at least 2.1 % (but usually between 3.0 to 4.5 %) carbon. The eutectic reaction gives a simultaneous precipitation of γ and cementite (in the metastable system) that is termed ledeburite. These alloys are suitable for casting due to the low melting point (eutectic temperature), their castability (ability to fill moulds), their high elastic modulus (rigidity) and their high damping capacity (ability to absorb vibration energy). Some grades of cast iron also have good machinability. There are four main grades of cast iron: white cast iron, grey cast iron, malleable cast iron, and ductile cast iron. These are obtained by variation of the heat treatment (the cooling rate) and the alloying elements added as shown in figure 6.8.

### 6.6.1 White Cast Iron

If the precipitation takes place within the metastable system (this is favoured by a relatively fast cooling rate, absence of silicon and nickel), then the structure consists of cementite and pearlite at room temperature (the austenite originally precipitated at 1146 °C undergoes the eutectoid reaction at 723 °C to form pearlite). Cast iron that contains most of its carbon in the form of cementite is termed white cast iron (a fractured surface has a bright appearance as a result of the presence of the brittle cementite). Due to the hard cementite; white cast iron is hard, brittle (percent elongation at fracture practically 0 %) and virtually non machineable. It may therefore be used only in those applications where wear resistance is the prime requirement e.g., in rollers of rolling mills.

As is the case with steels, white cast irons may be classified as hypoeutectic or hypereutectic. In hypereutectic white cast irons, the primary cementite (cementite precipitated before the eutectoid reaction) can only be retained if the material is quenched. The cementite that precipitates as part of ledeburite is termed ledeburitic or "free" cementite.

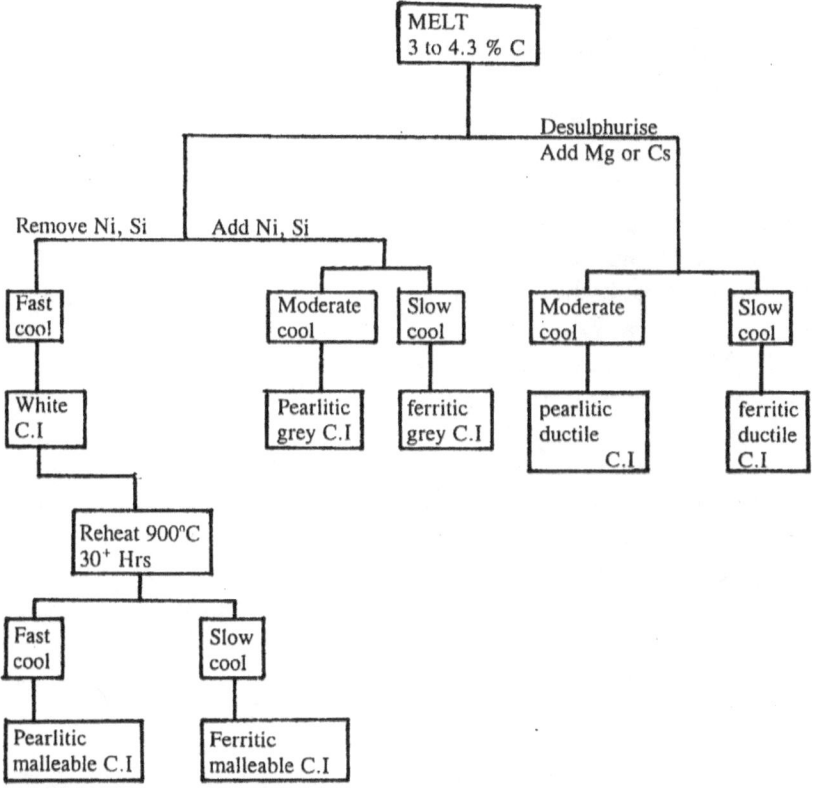

**Fig. 6.8** The various microstructures of cast iron

## 6.6.2 Grey Cast Iron

If the cooling rate is fairly slow and if appreciable amounts of silicon and/or nickel (2 to 5%) are present in the melt, the carbon will precipitate within the stable system as graphite in the form of graphite flakes. Cast iron that contains all its carbon as graphite flakes is termed grey cast iron (the graphite gives a grey appearance to a broken piece of this form of cast iron). The tendency to form graphite is also increased by increase in carbon content. The matrix may be either pearlitic or ferritic depending on the cooling rate.

Grey cast iron is weak and brittle in tension having a maximum elongation of only about 1 %. This results due to the tips of the graphite flakes which act as areas of stress concentration. For the same reason, grey cast iron has no notch sensitivity (i.e., the presence of notches does not severely reduce its strength since it has many notch like structures already).

However, they have high compressive strengths and ductility because the notch effect of the flakes does not affect compressive strength.

Grey cast iron has excellent vibration damping properties making it particularly suitable for base structures of machines subject to vibrations. They also have good wear resistance (making it suitable for cylinder block manufacture), and high fluidity (hence castability), making it possible to be cast into intricate shapes. Moreover, the flakes lead to easy chip breakage during machining accounting for its high machinability. Above all, grey cast iron is probably the cheapest structural material.

### 6.6.3   Malleable Cast Iron

Malleable cast iron is produced by heat treatment of white cast iron to reduce its brittleness and hardness. There are two industrial processes used for the heat treatment: In the "Blackheart" process, white cast iron is heated at about 900 °C for 2 to 3 days and cooled very slowly in a neutral atmosphere. The cementite breaks down to ferrite plus graphite in the form of rosettes. Thus, the final microstructure consists of graphite rosettes embedded in a ferrite matrix. In the "Whiteheart" process, the heating is done as above but in the presence of iron ore. Again, the cementite breaks down as above but near the surface, the ore oxides the graphite such that the structure at the surface becomes ferritic while the centre has a mixed structure consisting of ferrite, pearlite and graphite rosettes.

Malleable cast iron has adequate ductility and toughness while retaining adequate strength. It also has good machinability. Since the density of graphite is less than that of cementite, formation of malleable cast iron is accompanied by expansion. This is termed "growth" of cast iron. Typical uses include manufacture of connecting rods, gears, pipes, pipe fittings, etc.

### 6.6.4   Ductile or Nodular Cast Iron

When small amounts of magnesium or cesium are added to the melt before casting, formation of graphite flakes is suppressed and the graphite precipitates as spherical nodules. The result is ductile cast iron that is sometimes termed nodular or spheroidal cast iron. The matrix surrounding the nodules may be pearlitic or ferritic depending on the cooling rate (figure 6.8). Due to the elimination of the stress concentrating effect of the flakes, ductile irons are much stronger and more ductile than grey cast iron in tension. Indeed, the properties of ductile irons approach those of steels.

They are used for specialised casting e.g., pump bodies, crankshafts and gears.

## 6.6.5 Other Types of Cast Iron

Other commercially available forms of cast iron include: mottled iron, chilled iron and alloy cast iron. Mottled iron has a microstructure and properties between that of grey and white cast iron. Both graphite flakes and cementite are present in a pearlitic/ferritic matrix. Chilled irons are sand castings in which some parts are transformed to white iron by the use of metal chills in the mould. In alloy cast iron, other alloying elements are added to achieve specific properties. Chromium, nickel or molybdenum are added to increase the strength and wear resistance, 15-25 % nickel is added to produce wear resistant cast iron, etc. Martensitic cast iron may be obtained from grey or ductile cast iron by quenching from the austenite state so that martensite (see chapter nine) forms instead of pearlite. This is a heat treatment procedure designed to increase the hardness.

## PROBLEMS

6.1 The atomic radius of $\alpha$-Fe is 0.126 nm, while that of $\gamma$-Fe is 0.129 nm. Calculate the percentage changes in volume and linear dimensions that would be expected when $\alpha$-Fe changes to $\gamma$-Fe. At what temperature would you expect this change to occur and what is the phenomenon called? (Ans: -1.4 %, -0.47 %).

Hint on Q.6.1. Take 4 atoms of iron (1 unit cell FCC; 2 unit cells, BCC) and calculate volume for both structures. Then note that longitudinal strain $= 1/3$ volumetric strain.

6.2 A 0.07 % C steel is cooled under equilibrium conditions from 1600 °C to room temperature. Describe the phase changes that occur giving the approximate temperatures at which the changes take place.

6.3 What are the differences in structures and mechanical properties of grey, white, malleable and ductile cast irons?

6.4 Sketch and label the microstructures of 0.2 % C; 0.8 % C and 1.4 % C steels at room temperature.

6.5 Plot a phase transformation diagram for 0.6%C steel.

6.6 A malleable cast iron containing 3.0 % C was initially solidified as γ + cementite. It was then re-heated to dissociate the cementite. What is the maximum possible volume percentage of graphite at 800 °C and 722 °C? The densities of iron and graphite are 7600 kg/m³ and 2000 kg/m³ respectively. (Ans: 7.2 % graphite; 10.5 % graphite).

6.7 Draw the Fe-C diagram approximately to scale and explain the changes in structure which occur when a 0.2 % C alloy is cooled from 1600 °C to room temperature.

6.8. For a plain carbon steel containing 0.3 %C:
(i) Sketch the likely crystal structures at 1000 °C, 724 °C and 722 °C. Assume equilibrium conditions.
(ii) Calculate the quantities of pro-eutectoid ferrite, pearlite, eutectoid ferrite and eutectoid cementite present in 1 kg of the alloy at 722 °C. (Ans: 641 g, 359 g, 316 g, and 42 g).

6.9 Sketch schematically the change in volume (with temperature), which occurs when iron is heated from room temperature to melting. Give values of the important temperatures and explain the shape of the curve.
What changes in volume and linear dimensions occur when ferrite changes to austenite? (Atomic radii: α: 0.126 nm, γ: 0.129 nm.) (Ans: -1.4 %, -0.47 %.)

6.10 (a) Using iron as an example, explain what is meant by allotropy and hence explain the existence of ferrite and austenite.
(b) A steel containing 0.007 % C is cooled under equilibrium conditions from 1600 °C to room temperature. Describe all the changes that occur in crystal structure giving the approximate temperatures at which the changes in structure take place.
How much pearlite is present in 500 g of the above alloy just below the eutectoid temperatures? (Ans: 0 g.)

6.12 Draw the iron-cementite diagram to scale and determine, for 900 g of a 1.35 % C steel and a 0.47 % C steel (at 722 °C):
(i) The quantity of pro-eutectoid cementite or ferrite (Ans: 84.3 g; 380.8 g).

(ii) The quantity of pearlite (Ans: 815.7 g; 519.2 g).
(iii) The quantity of eutectoid ferrite (Ans: 717.8 g, 456.9 g).
(iv) The quantity of eutectoid cementite (Ans: 97.9 g; 62.3 g).

6.13 Sketch, approximately to scale, the iron-carbon diagram up to 2.1 % C (the "steel" part) showing all important phases and temperatures. From the diagram or otherwise, determine the quantities of (i) pro-eutectoid cementite, (ii) pearlite, (iii) eutectoid ferrite, (iv) eutectoid cementite, present in 600 g of a 1.3 % C steel which is at 722 °C. (Ans: 51.1 g, 548.9 g, 483 g, 65.9 g.)

6.14 Show in a sketch how the strength of plain carbon steels, which are heated and quenched, vary with carbon content and explain the shape of the curve.

6.15 Write equations for the eutectic, peritectic and eutectoid reactions, and sketch a familiar equilibrium diagram in which all the three reactions take place, naming the phases and showing the appropriate invariant points.

## Further Reading

Van Vlank, L.W. Materials for Engineering, Addison-Wesley, Reading, 1982.

Callister, W. D. Materials Science and Engineering, 3rd ed., John Wiley & Sons, New York, 1994.

Rollason, E. Metallurgy for Engineers, 4th ed., ELBS & Edward Arnold, Norwich, 1973.

# CHAPTER SEVEN

# ALUMINIUM AND ITS ALLOYS

## 7.1 PROPERTIES OF ALUMINIUM AND ITS ALLOYS
### 7.1.1 Advantageous Properties

Aluminium and its alloys have 5 major properties that make them useful for engineering purposes:

(1) Low specific gravity. Pure aluminium has a specific gravity of 2.7 as compared to 8.8 for copper and 7.8 for iron and steel. It therefore becomes indispensable in applications where high strength to weight ratio is important. Its alloys are hence extensively used in the aerospace industry, in automobile manufacture (for components like connecting rods and pistons) and for structures like ladders where weight is a prime consideration.

(2) Good electrical conductivity. Pure aluminium has a conductivity about 60 % that of copper. But its light weight means it has a much higher specific conductivity. It is therefore exclusively used for overhead transmission cables (with steel reinforcement for strength).

(3) Good thermal conductivity. Aluminium and its alloys have high thermal conductivities and are hence used for heat exchanger components and in machine parts where rapid dissipation of heat is important e.g., motor engine cylinder heads, gear boxes, etc.

(4) Good corrosion resistance. Aluminium is resistant to corrosion due to the formation of a thin impervious layer of $Al_2O_3$ that forms on any aluminium article when exposed to oxygen (air). The high affinity of aluminium for oxygen (see point 5 hereafter) ensures the immediate formation of this layer, which then prevents further contact between the aluminium and oxygen.

When formed naturally, this layer is very thin. To improve corrosion resistance, the layer is made thicker in a process known as anodizing. In this process, the article is made an anode in an electrolytic cell containing oxalic acid (chromic and sulphuric acids are also used), the cathode being lead. When current is passed, oxygen is evolved at the anode and combines with the aluminium. This increases the thickness of the layer to approximately 0.015 mm (without anodizing, the layer is about $10^{-5}$ mm), greatly enhancing the corrosion resistance. This property makes aluminium and its alloys useful in several situations:

(i) In chemical plants handling nitric acid and other corrosive chemicals.

(ii) For cooking pans and other holloware.

(iii) As aluminium foil for food packaging (e.g., in oven foils, in cigarette packaging, etc.)

(iv) As aluminium paint.

(v) In building and marine applications.

The corrosion resistance is best in pure aluminium, followed by alloys of magnesium, manganese, magnesium-silicon, silicon and worst in copper alloys.

5. Affinity for oxygen. As noted in point 4, aluminium has a high affinity for oxygen. This is put to industrial use in:

(i) Steel manufacture: Aluminium is used as a de-oxidant in steels. Oxygen is a major impurity in steel and hence in steel manufacture, aluminium is added to the steel melt. Its high affinity for oxygen ensures that it forms oxides thus removing the impurity. Steel so treated is said to be "killed".

(ii) Thermit welding: The process of thermit welding is used to repair large iron or steel castings. A mould is made around the part to be repaired and filled with thermit powder (a mixture of powdered aluminium and iron oxide). When the mixture is fired, the following reaction takes place:

$$Fe_2O_3 + 2Al \longrightarrow Al_2O_3 + 2\,Fe \tag{7.1}$$

This reaction is exothermic and the heat generated melts the iron produced. The molten iron flows into the broken part and fusses the two parts together, thus producing the weld.

In addition, some alloys of aluminium are weldable, while others have good formability. They are also non-toxic and hence may be used as containers for food.

## 7.1.2 Disadvantageous Properties

The following properties of aluminium work to its disadvantage:

(1) High affinity for oxygen. Having a higher affinity for oxygen than the common reducing agents (e.g., carbon) means that aluminium cannot be economically extracted from its ore by reduction. Neither can blowing air into the melt purify it. If this is attempted, aluminium will be oxidized before the impurities. Thus, aluminium has to be extracted and purified by the expensive method of electrolysis. The process used is termed the Hall process and involves dissolving the ore (hydrated $Al_2O_3$ known as bauxite)

in molten cryolite ($Na_3AlF_6$). The solution is used as an electrolyte in an electrolytic cell lined with carbon that acts as the cathode. The anode is a carbon rod dipped in the solution. Aluminium ions are deposited at the cathode and since the melting temperature of cryolite (about 950 °C) is higher than that of aluminium (660 °C), the aluminium is molten. The molten salt is lighter than the metal and hence covers the same thus protecting it from contamination.

(2) High thermal expansion. Aluminium has a high coefficient of thermal expansion. Due to this fact, adequate allowance has to be made for this expansion when aluminium is used in high temperature applications (e.g., heat exchangers). If this is not done, high thermal stresses will be produced.

## 7.2 ALUMINIUM ALLOYS

Pure annealed aluminium is relatively weak, having an ultimate tensile strength of about 45 MPa. It can therefore not be used for structural purposes. For structural applications, aluminium is alloyed with one or more of the following elements: copper, magnesium, manganese, zinc, silicon and lithium. The resulting alloys may be available as wrought products i.e., worked by processes like rolling, extrusion, etc., or as castings. Furthermore, both cast and wrought alloys may be heat treatable i.e., may have their properties improved by a series of heating and cooling operations, or non-heat treatable. The properties of the non-heat treatable alloys may still be improved by work or strain hardening.

### 7.2.1 Designation of Aluminium Alloys

Several standardization bodies and professional associations have developed methods for designating aluminium alloys. One of the most often used procedures is that developed by the Aluminium Association, AA, which is briefly outlined hereafter: (By the time of writing this book, a Kenya Standard on the designation of aluminium alloys had not been developed.)

Wrought alloys are designated by a four-digit code, the first of which identifiers the major alloying element(s) as follows:

| | |
|---|---|
| Commercially pure aluminium i.e., no alloying | 1XXX |
| Copper | 2XXX |
| Manganese | 3XXX |
| Silicon | 4XXX |

| Magnesium | 5XXX |
|---|---|
| Magnesium and silicon | 6XXX |
| Zinc and magnesium | 7XXX |
| Other (e.g., lithium) | 8XXX |

In addition to the alloying elements, the designation also shows the treatment to which the alloy has been given. This is shown by a letter (preceded by a hyphen) following the four digits. The letters are: O for annealing, H for strain hardening, and T for heat treatment. A digit that further specifies the nature of heat treatment may follow this. For example, the digit 3 stands for natural aging after stretching, digit 4 for natural aging without stretching, digit 6 for artificial aging without prior stretching, digit 8 for artificial aging after stretching, and digit 7 for over aging. Some of these terms will be expounded on in the following sections. As an example, the alloy designated AA 2024-T8 is a wrought aluminium alloy which has copper as the major alloying element, which has been artificially aged after stretching, etc. The AA stands for Aluminium Association.

Of the alloy series considered, alloys in the 2XXX, 6XXX, 7XXX, and some in the 8XXX series are heat treatable while the rest are not.

Cast alloys on the other hand are given a three-digit designation sometimes followed by a dot and another digit. A preceding letter indicates modifications of existing alloys. Again, the leading digit identifies the major alloying element as follows:

| Copper | 2XX (or 2XX.X) |
|---|---|
| Silicon plus copper and/or magnesium | 3XX |
| Silicon | 4XX |
| Magnesium | 5XX |
| Zinc | 7XX |
| Tin | 8XX |

Examples are A206 and B356.2. The cast alloys in the 2XX, and 3XX series are heat treatable.

## 7.3  HEAT TREATMENT OF ALUMINIUM ALLOYS

### 7.3.1  General Overview

The term heat treatment refers to a series of heating and cooling operations designed to alter the properties of a material. The heat treatment procedure used for aluminium alloys is termed precipitation hardening or age hardening, and occurs in binary systems where there is a sharp decrease in the solubility of one element in the other as the temperature is decreased. An example is the binary Al-Cu phase system. A portion of the phase diagram is shown in figure 7.1. If equilibrium is attained, any alloy containing less than the eutectic composition of copper consists of two phases at room temperature: $\alpha$ (solution of Cu in Al) and $\theta$ (the intermetallic compound $CuAl_2$). $\alpha$ is referred to as the matrix while $\theta$ is the precipitate. Under equilibrium conditions, the precipitates occur as large particles that are incoherent (this term will be expounded on shortly) with the matrix. This microstructure is ineffective in blocking dislocation motion and hence the resulting material is very weak. From figure 7.1 however, it may be noted that an alloy with 4.5 % Cu, when heated to a temperature above 540 ºC consists of the single phase $\alpha$, i.e., all the copper is in solution. This procedure (i.e., heating to a temperature where the second phase is fully dissolved) is termed solution heat treatment. If the solution is now quenched in water, the cooling rate is too fast for the copper to diffuse out of solution and form the equilibrium phase, $\theta$. This is because formation of $\theta$ requires diffusion of atoms to new sites, a process that requires time. A supersaturated solution of copper in aluminium results. If this is allowed to stand at room temperature, it is noted that the hardness of the material increases with time. The material is said to "age". If this takes place at room temperature, it is termed natural aging. In the 4.5 % alloy, maximum hardness is achieved in about 10 days. The rate of hardening is increased if the material is heated to a temperature between 130 ºC and 190 ºC after quenching. Heating the material to an elevated temperature and holding at this temperature is termed precipitation heat treatment or artificial aging. In artificial aging, the hardness falls if the material is aged for too long. The reduction of hardness with time is called over aging or reversion. It should be clear from this that precipitation hardening requires careful control of time and temperature. The mechanisms responsible for the changes described so far are considered hereafter.

## 7.3.2 Coherent Versus Incoherent Precipitation.

When two solid phases co-exist, there must be an interfacial boundary between them. This boundary is said to be fully coherent if there is a one to one continuity in the atomic planes of the two phases on all boundaries, as shown in figure 7.2 (a). Full coherency is only possible if the crystal systems of the matrix and precipitate are the same. However, it is unlikely that the atomic spacing of both is the same so that the area around the precipitate is strained. These strains are termed coherency strains. There is strain energy associated with the strains such that movement of dislocations through the precipitate is restricted. This accounts for the strengthening effect of precipitates. Further contribution to the interfacial energy comes from the differences in chemical composition between the precipitate and matrix (chemical interfacial energy).

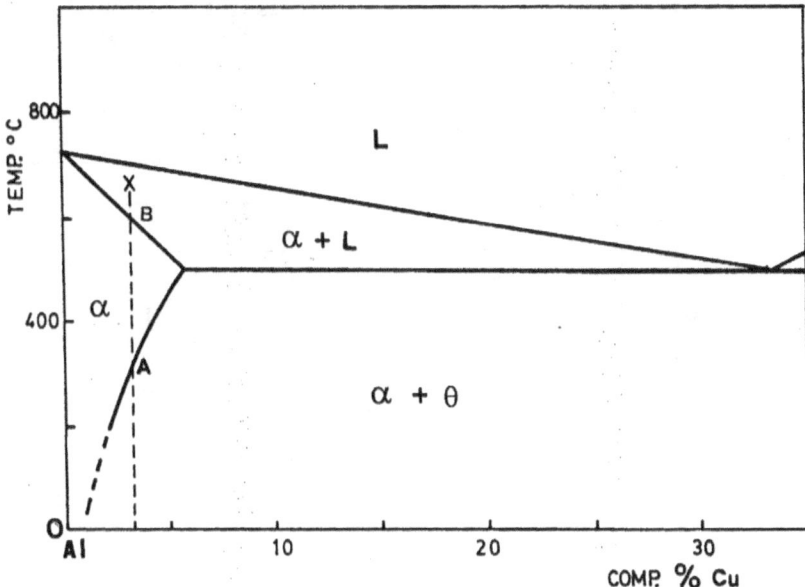

**Fig. 7.1** A portion of the Al-Cu phase diagram

The interfacial boundary is said to be semi-coherent if the planes correspond in one direction but not in the other, or if the differences in atomic spacing are large, such that the mismatch is taken up by periodic edge dislocations. An example of a semi coherent interface is illustrated in figure 7.2 (b). Semi coherent interfaces will result if the crystal structures of the precipitate and matrix are different, but the two have similar planes. For example, the {111} planes in an FCC matrix will match with the {0001} planes of a HCP precipitate since both planes are hexagonally close packed. However, no other planes match meaning that there will be no one to one correspondence in the other interfaces. For this reason, semi coherent precipitates tend to be disc shaped or plate like. The plane in the matrix that corresponds to a plane in the precipitate is termed the habit plane. In the case of a HCP precipitate in an FCC matrix for example, the habit plane is {111}.

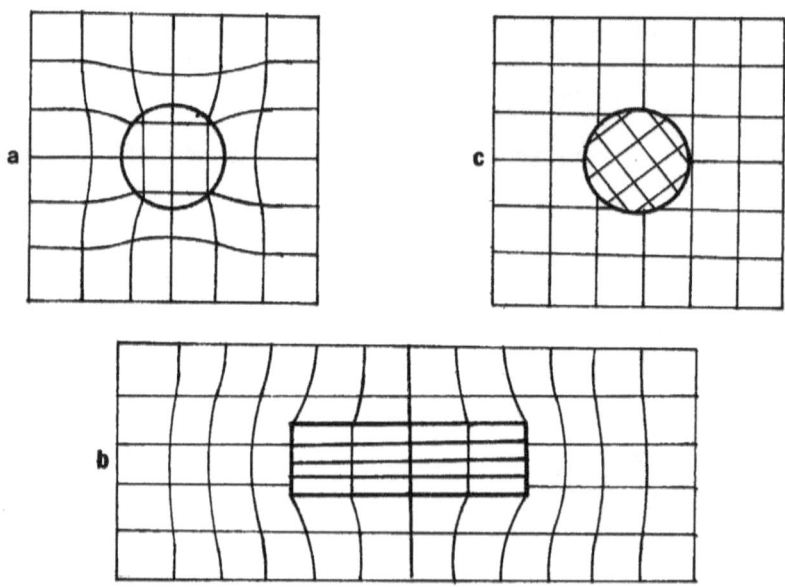

**Fig. 7.2** Coherency of precipitates: (a) coherent (b) semi-coherent (c) incoherent

Finally, the interface is said to be incoherent if there is no correspondence of planes in any direction (figure 7.2 (c)). Incoherent precipitates tend to be spherical in shape. As expected, the larger the precipitate, the less the chances of it being coherent. Incoherent precipitates introduce no coherency strains hence contribute little to strengthening.

### 7.3.3 Nature of Precipitates in Al-Cu Alloys.

The supersaturated solution resulting from quenching after solution heat treatment is unstable and will tend to change to the equilibrium phases with time. The lower the temperature, the larger the driving force for the change. However, lower temperature means lower mobility of atoms hence lower energy available to activate the process. Since the activation energy required for the direct formation of θ is large, the transformation takes place in stages. In other words, several transition phases, whose formation requires less activation energy, form before the equilibrium phases. This is illustrated schematically in figure 7.3.

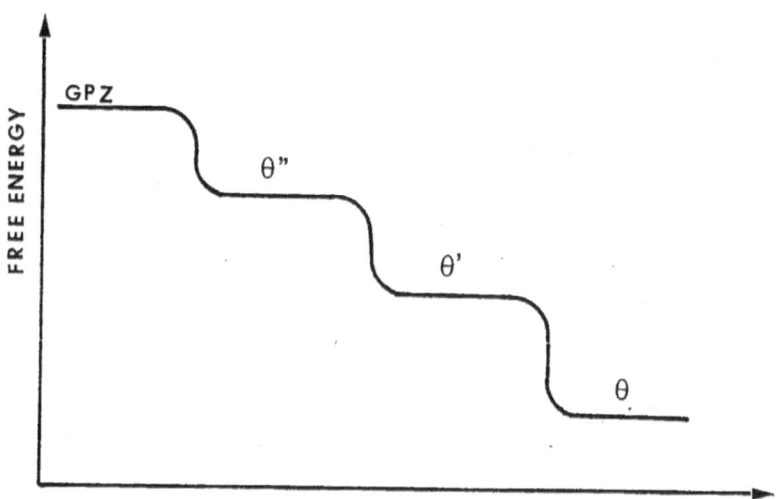

**Fig. 7.3** Schematic showing total free energy of transition phases in the Al-Cu system

In the Al-Cu system, natural aging starts when the excess copper diffuses to copper rich clusters termed Guinier-Preston or G-P zones. The zones are disc shaped and fully coherent with the $\alpha$-matrix. They have a {001}-habit plane, are about two atom layers thick, and reach a diameter of about 10 nm. As their size increases, their effectiveness in blocking dislocations increases. This explains the increase in hardness that constitutes age hardening. However, past a certain size, it becomes energetically favourable for the zones to form their own crystal structure. This new phase has the approximate formula $CuAl_2$, and a tetragonal crystal structure (lattice parameters 0.404 nm x 0.404 nm x 0.768 nm), which may be thought of as a distorted FCC crystal. The matrix is, of course, FCC with a lattice parameter of 0.404 nm. The closeness in the crystal structures explains the lower activation energy required for this change. The phase, which is itself metastable, is termed $\theta''$, or G-P zone II. It is fully coherent with the matrix and this, together with their optimum size (about 100 nm) leads to maximum strengthening effect. In natural aging, no further transformation takes place and the hardness achieved is retained as shown in figure 7.4.

If aging is carried out at a higher temperature, e.g., 130 °C, there is formation of another transition phase, $\theta'$, which has a formula and crystal structure similar to $\theta''$. However, the lattice parameters are 0.404 nm x 0.404 nm x 0.580 nm, making it incoherent with the matrix on the (100) and (010) planes. The (001) planes are however, identical with the {001} planes of the matrix. Hence, $\theta'$ is a semi coherent precipitate. If the aging time is increased, $\theta'$ grows in size and loses coherency, and hence its ability to block dislocation motion. The hardness therefore decreases accounting for the over aging effect noted earlier. Further aging at this temperature leads to the formation of the equilibrium phase $\theta$. This has the same formula as $\theta'$ but the crystal structure is body centred tetragonal (lattice parameters 0.607 nm x 0.607 nm x 0.487 nm). It is totally incoherent with the matrix. Moreover, it occurs as large, widely spaced conglomerates, which are totally unable to block dislocation motion. (Though the dislocations cannot cut the precipitates, they still pass them by looping around.)

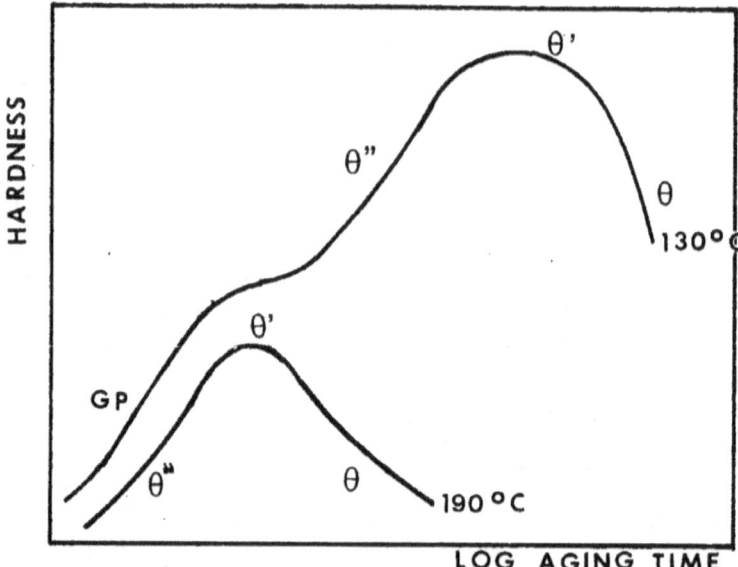

**Fig. 7.4** Schematic variation of hardness with time during the aging of an Al--4.5 % Cu alloy.

A similar sequence of events takes place when aging takes place at 190 °C with the difference that the changes take place much faster and that no G-P zones are formed (the dissolution of G-P zones and other precipitates at higher temperature is termed reversion). The highest hardness is achieved when the microstructure consists of a mixture of $\theta''$ and $\theta'$. Thus the precipitation sequence in Al-Cu alloys may be written as:

$$\alpha \,(\text{ss}) \rightarrow \alpha_1 + \text{GPZ} \rightarrow \alpha_2 + \theta'' \rightarrow \alpha_3 + \theta' \rightarrow \alpha_4 + \theta \qquad (7.2)$$

In equation 7.2, ss stands for supersaturated solution, and $\alpha_i$ refers to the solid solution in equilibrium with each metastable phase. These changes are summarized in figure 7.4. The higher hardness achieved through artificial aging at 130 ºC compared to 190 ºC results from the additional fact that precipitation reactions at higher temperatures result in faster precipitation rates and hence coarser precipitates.

Similar precipitation reactions occur in other heat treatable Al-X binary alloys and in ternary alloys. In the ternary Al-Cu-Mg system for example, the metastable phase providing hardening is S' ($Al_2CuMg$). In Al-Cu-Li, the strengthening phases are $T_1$ ($Al_2CuLi$); $\theta'$ ($Al_2Cu$); and $\delta'$ ($Al_3Li$). In the 6XXX (Al-Mg-Si) series the precipitate is $\beta'$ ($Mg_2Si$), while in the Al-Mg-Zn (7XXX) series, the precipitate is $\eta'$ ($MgZn_2$). The heat treatable alloys in the 8XXX series derive their strength from S', $\delta'$ ($Al_3Li$) among other precipitates.

If the material is plastically deformed prior to aging, higher hardness is achieved. The reason for this is that the deformation produces a multiplication of dislocations which act as nucleation sites for the precipitates. The many nucleation sites result in finer precipitation and hence higher hardness. With appropriate combination of precipitation hardening and plastic deformation, aluminium alloys with tensile strengths approaching 1000 MPa have been produced. This is a momentous achievement considering that the strength of pure aluminium is a mere 45 MPa.

## 7.4 MODIFICATION IN CAST ALLOYS

A process known as modification may improve the strength of the non-heat treatable 4XX.X cast alloys. This may be understood by considering the part of the Al-Si phase diagram shown in figure 7.5. The eutectic temperature is shown to be 577 ºC with a corresponding composition of 11.6 weight percent silicon. Castings resulting from this equilibrium solidification tend to be coarse grained and weak. If a small amount of sodium is added to the melt, nucleation is delayed due to formation of a thin layer of sodium on any nascent nuclei. This allows for cooling of the melt to a temperature below the solidification or melting temperature. When solidification starts at 564 ºC, the under cooling is large, resulting in the formation of a large number of nuclei simultaneously. The end result is a fine-grained casting with improved mechanical properties. The process also shifts the eutectic composition to 14 % silicon.

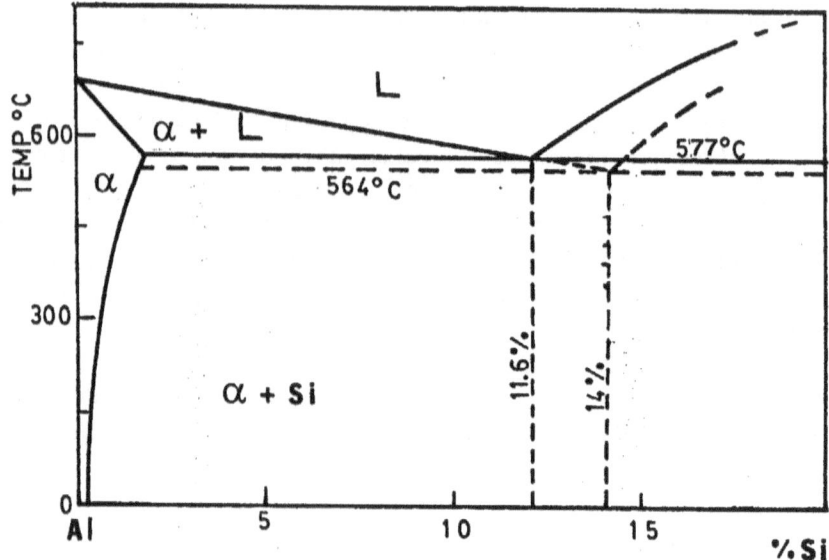

**Fig. 7.5** Modification of Al-Si cast alloys

## 7.5 ALUMINIUM LITHIUM ALLOYS

A new class of aluminium alloys has gained prominence towards the end of the last millennium due to the discovery that modest addition of lithium to aluminium result in an increase of up to 10 % in the value of Young's modulus accompanied by a reasonable decrease in the density. The resulting alloys are termed aluminium-lithium alloys (though nearly all of them are ternary and higher order alloys). Nearly all the alloys developed so far are wrought heat treatable alloys. Strengthening is provided by $\delta'$, $\theta'$, S' and $T_1$ whose stoichiometry have already been given previously. The high modulus, low weight alloys are of particular interest to aircraft and aerospace manufactures. Moreover, certain versions are weldable and may soon be the major material for manufacture of cryogenic tanks. However, some of these alloys suffer from low ductility and toughness, in addition to anisotropy. Commercially available aluminium-lithium alloys include AA 2090, AA 2091, AA 2095, AA 8090 and AA 8091.

## 7.6 USES OF ALUMINIUM ALLOYS

The major uses of the various series of aluminium alloys are summarized in Table 7.1.

**Table 7.1** Major uses of aluminium alloys

| 1XXX | Aluminium foil, in capacitors, sheet work, tubing, electrical conductors |
|---|---|
| 2XXX | Aircraft, aerospace, welded cryogenic tanks, cylinder heads, auto pistons, wheels |
| 3XXX | Cooking utensils, sheet, buildings (siding, gutters) |
| 5XXX | Bus/truck bodies, marine/chemical storage tanks, utensils, buildings, appliances (e.g., fridges), missiles, amour plate |
| 6XXX | Auto body sheets, wires, rods, sheets, plates, marine structures, railroad cars, furniture |
| 7XXX | Aircraft structures, truck bodies, missiles, etc. |
| 8XXX | Aircraft, spacecraft |
| 2XX | General sand castings, valve bodies, cylinder heads, pistons, pump bodies |
| 3XX | Engine parts, grilles, cylinder heads, meter housings, aircraft castings, pistons, crank cases, aircraft wheels, fan blades, pulley sheaves, valve bodies, axle housings, etc. |
| 4XX | Large castings with thin sections, instrument casing, typewriter frames, etc. |

## PROBLEMS

7.1 Describe two industrial processes, which rely on the high affinity of aluminium for oxygen. Give reasons why aluminium is resistant to corrosion despite this high affinity and expound on the main disadvantage of this property.

7.2 What are the main uses of aluminium-silicon cast alloys? How can the structure and properties of these alloys be changed by "modification"? Illustrate your answer by referring to the Al-Si phase diagram.

7.3 Describe the structural changes that occur during the precipitation heat treatment of an aluminium-copper alloy containing 4% Cu. What precautions are necessary if this process is to be accelerated by heating and why?

7.4 (a) Given that one property of aluminium, which accounts for its widespread use, is its corrosion resistance, explain the basis of this resistance and state 5 major industrial applications where this property is put to use. How can the resistance be improved?

(b) Explain how the heat treatment of wrought aluminium alloys is carried out.

(c) Briefly describe four other uses of aluminium and its alloys.

7.5 An aluminium-copper alloy is being precipitation hardened. After quenching, it is aged at room temperature, 100 °C, 200 °C and 250 °C. Sketch, on the same graph, the variation of hardness against time at each temperature and explain the shape of the curves.

7.6 Given that aluminium has a high affinity for oxygen:

(i) Describe two industrial processes where this property is put to use.

(ii) What is the main disadvantage of this property?

(iii) How can you explain that, despite this affinity, aluminium is resistant to corrosion? How can this corrosion resistance be improved?

7.7 Explain, with reference to the copper-aluminium alloy system, what is meant by precipitation hardening.

7.8 (a) State the main uses of aluminium and the corresponding properties that make it suitable for these uses.

(b) Using a 3% Cu alloy of aluminium and copper as an example, explain what is meant by precipitation hardening. What precautions would you take and why if ageing of the alloy is accelerated by heating?

## Further Reading

Higgins, R.A. Engineering Metallurgy Part I, Hodder & Stoughton, London, 1980.

Rollason, E.C. Metallurgy for Engineers, 4th ed. Edward Arnold, Norwich,

1973.

Hatch, J.E. (ed.) <u>Aluminium, Properties and Physical Metallurgy,</u> ASM, Metals Park, Ohio, 1984.

Schaffer, J. P. et al. <u>The Science and Design of Engineering Materials</u>, Irwin, Chicago, 1995.

# CHAPTER EIGHT

# OTHER NON-FERROUS ALLOYS

## 8.1 COPPER AND ITS ALLOYS
### 8.1.1 Introduction

The most important property of pure copper is its good electrical conductivity, which is second only to that of silver. The conductivity is at its best in the pure annealed metal and is greatly affected by impurities. The effect of impurities on the conductivity is shown schematically in figure 8.1 from which it can be seen that the most profound effect is caused by phosphorous and silicon. However, cadmium reduces the conductivity only nominally and is usually added to copper for telephone wires to improve the strength.

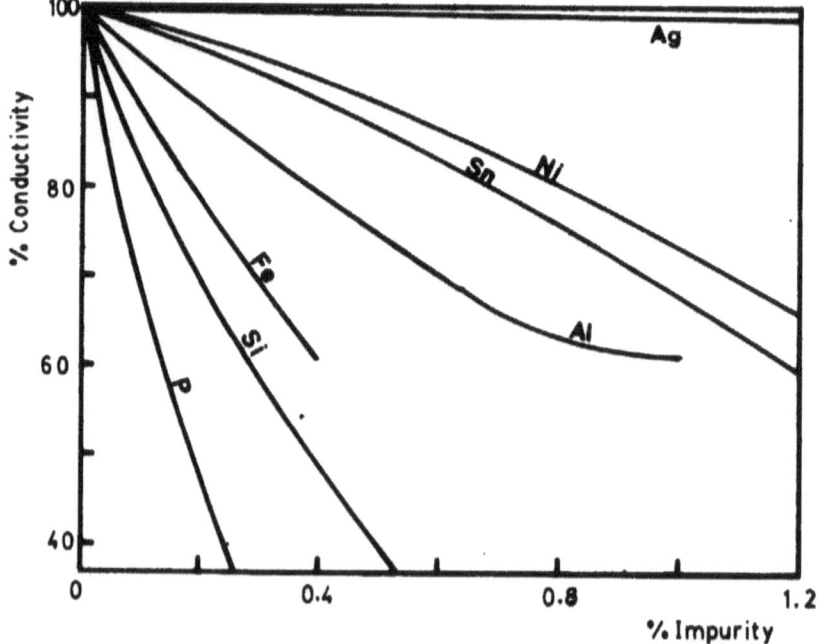

**Fig. 8.1** The effect of alloying elements on the conductivity of copper

Nearly all alloying elements increase the strength of copper. The only exceptions are bismuth and antimony, which segregate to the grain boundaries thereby causing britleness.

Copper forms two commercially important series of alloys namely brasses and bronzes. <u>Brasses</u> are principally alloys of copper and zinc containing up to 45 % Zn. They are important due to the following properties: a wide range of mechanical properties, they are soft and easy to work, have a pleasant appearance and are largely resistant to corrosion.

Figure 8.2 shows a portion of the Cu/Zn equilibrium diagram. From the diagram it can be seen that brasses can be divided into two major groups: α brasses and α–β brasses. α brasses contain up to 38% zinc and consist of a single-phase α at room temperature. α, being a solid solution, is soft and ductile. The highest ductility (approximately 69 %) occurs at about 30 % Zn. α-brasses are therefore suitable for cold working and are cold rolled into sheets, rods, tubes and wires. The most popular α-brass is the 70/30 brass (with 70 % copper, 30 % zinc). Cold working increases the strengths of α brasses.

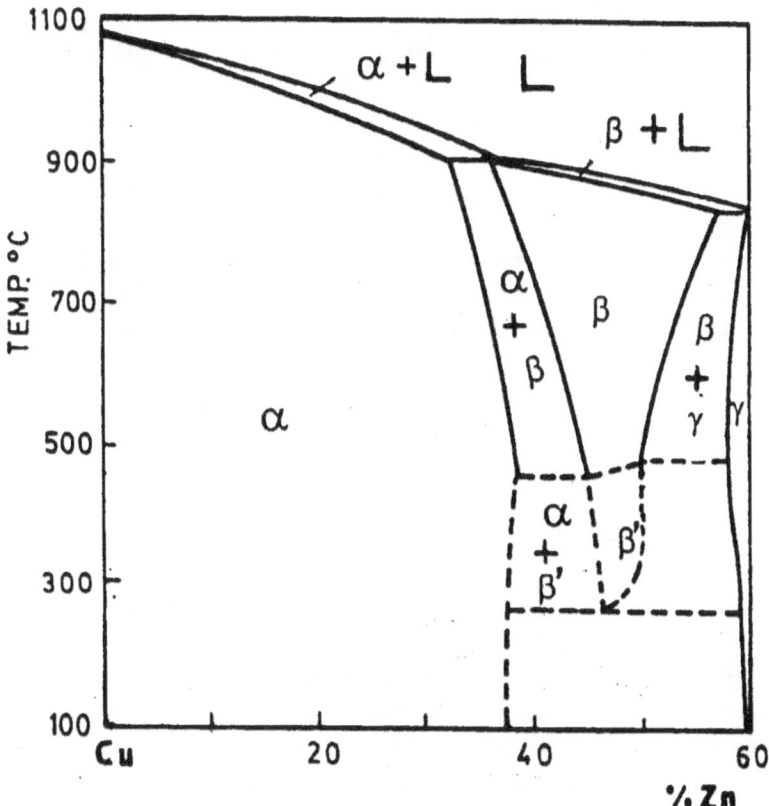

**Fig. 8.2** The copper rich portion of the copper-zinc phase diagram

α–β brasses on the other hand contain between 38 and 46 % copper. They consist of two phases α and β at room temperature. They have relatively high strength but much reduced ductility compared to α-brasses. Being multi-phase and having low ductility at room temperature, α–β brasses cannot be easily cold worked. They are usually hot worked above 800 °C where they consist of the single-phase β.

α–β brasses are suitable for hot working (extrusion, hot rolling, etc.) The most abundant commercial variety is the 60/40 brass.
There is a rapid fall in both the strength and the ductility for alloys containing more than 45 % Zn due to the appearance of the brittle phase γ.

Other commercially available brasses include: leaden brass or free machining brass. This is brass to which 1.5 to 3.5 % lead has been added to improve the machinability. Lead is insoluble in copper and collects at grain boundaries making it easier for chips to break during machining; high tensile brass is α–β brass to which one or more of the following elements have been added to improve the strength: iron, aluminium, manganese, tin, nickel. High tensile brass can have ultimate tensile strengths up to 740 MPa.

Bronzes The alloy most often referred to as bronze is tin bronze. This alloy contains about 10 % tin. The relevant portion of the equilibrium diagram is shown in figure 8.3, which indicates that the equilibrium phases at room temperature should be α + ε. However, diffusion is so slow that the crystal structure remains α at room temperature. α is tough, ductile, and cold workable. Tin bronze, which is the most abundant bronze commercially, contains about 7 % Sn and is supplied as rolled sheets or rods.

Tin bronzes are also supplied as castings containing about 18 % Sn. These alloys however have a problem of coring that may make the brittle phase, δ, appear at phase boundaries. To be useful, these castings need prolonged annealing at about 700 °C to form α, which is then cold worked. The un-annealed cast alloys are used for bearings due to their high wear and shock resistance.
There are four other commercially available bronzes containing elements other than (or in addition to) tin:

1. Phosphor bronze. These contain 0.1 to 1.0 % phosphorous in addition to copper and tin. Phosphorus increases the strength and improves the corrosion resistance. They may be supplied wrought (as wires or rods) or cast. They are used for bearings requiring high strength due to their low coefficient of friction.

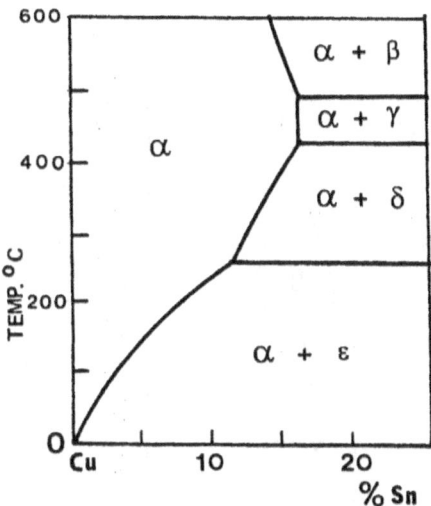

**Fig. 8.3** Portion of the copper-tin phase diagram

2. Bronzes containing zinc. Contain about 3 % Sn together with approximately 2.5 % Zn. The alloy is used making "copper" coins. When cast (10 % Sn plus 2 % Zn), zinc bronzes are used for corrosion resistant castings and for military decoration "gunmetal".

3. Leaden bronzes. These are tin bronzes containing about 2 % lead, which improves their machinability. The bronzes have good thermal conductivity and are used in high-speed bearings e.g., aircraft and automobile crankshaft bearings.

4. Aluminium bronze. These are alloys of copper and aluminium. The relevant part of the equilibrium diagram is shown in figure 8.4 and shows a eutectoid reaction at 565 °C and 11.8 % Al). Due to this eutectoid reaction, heat treatment, similar to formation of martensite in steels, is possible. A 10 % Al alloy for example consists of β at 900 °C. This can be water quenched to produce β', a non-equilibrium phase which is hard and brittle.

In addition to their capability to undergo heat treatment, aluminium bronzes have the following useful properties:

(a) They retain their strength at high temperatures.
(b) They have good resistance to corrosion that is retained at high temperatures.
(c) They have good wear resistance.
(d) They have a pleasant gold like appearance and are hence used in imitation jewellery.

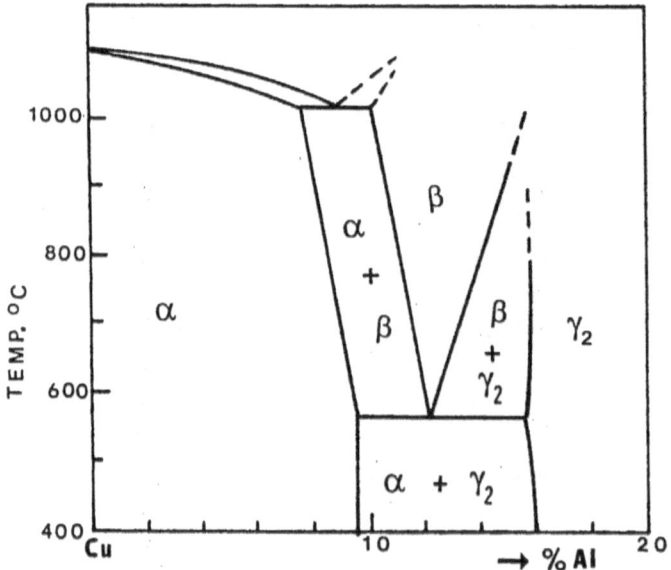

**Fig. 8.4** Portion of the copper rich part of the Al-Cu phase diagram

Looking at figure 8.4, it is evident that there are two series of alloys:
α alloys. These usually contain 4 to 7 % aluminium with the microstructure consisting of the single phase, α. They are used in manufacture of condenser tubing and for imitation jewellery. Being single phase, they are cold workable, have moderate strength and good corrosion resistance.

α–$\gamma_2$ alloys: Contain 7 to 12 % Al and consist of the two phases α and $\gamma_2$ at room temperature. Like steels, they can be hot worked (forged/hot rolled) after heating to the single phase, β. They are used in chemical engineering plants especially those exposed to high temperatures. This is mainly due to their corrosion resistance at high temperatures.

Aluminium bronzes can also be cast (9 to 12 % Al). The castings are used in marine applications for pump rods, propellers, etc., in automobile engines for valve seats, spark plug bodies, and for the manufacture of heavy-duty bearings and gears.

Cu-Ni alloys: In presence of tin, copper--nickel alloys have a silvery appearance and are used in the manufacture of "silver" coins. They are also used as electrical resistors and for high strength springs. They can attain strengths up to 900 MPa. Copper and nickel show complete solubility in the solid state and hence from a whole series of alloys termed "cupro-nickels". These are used for condenser tubes and, with iron, in the manufature of thermocouples.

## 8.2 TITANIUM ALLOYS

Titanium has a density in between those of steel and aluminium (specific gravity of 4.54) and a relatively high melting point (1668 °C). Its alloys have a high strength to weight ratio and maintain their properties up to fairly high temperatures (about 550 °C). They also have excellent corrosion resistance. The metal itself is extracted from the mineral rutile ($TiO_2$) and has a temperature based polymorphism. Up to 883 °C, the crystal structure is hexagonal close packed. This polymorph is termed α-titanium. Beyond this temperature, the crystal structure changes to body centered cubic, a polymorph termed β-titanium.

Despite of its advantages, titanium is used only for specialized application due to its high cost. This results mainly from its highly reactive nature and its affinity for the non-metals (oxygen, hydrogen, nitrogen and carbon). As such titanium is very difficult to extract. Moreover, its fabrication has to be done in vacuum.

Titanium forms two main alloy systems: the α-stabilized and the β-stabilized alloy systems.

### 8.2.1 α-Stabilized Ti-Alloy Systems

This series results when titanium forms alloys with elements that stabilize the α-polymorph. These elements (the main one of commercial interest being aluminium), dissolve preferentially in α-Ti. The resulting phase diagram has the α-field expanded due to a rise in the α to β transition temperature as shown in figure 8.5.

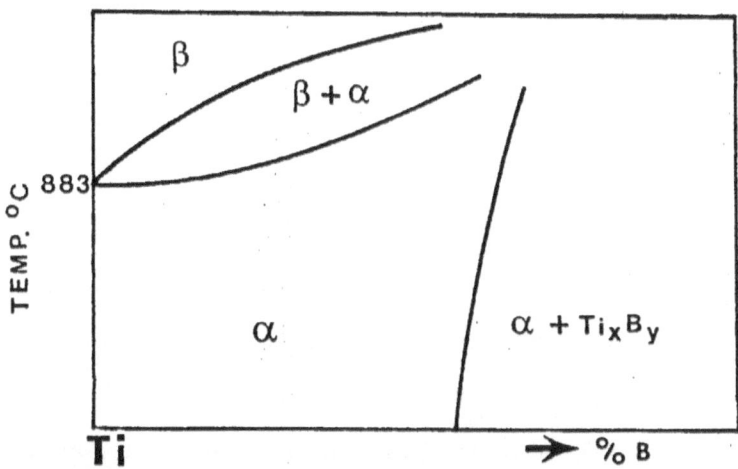

**Fig. 8.5** Phase diagram between Ti and an α-stabilizer

## 8.2.2 β-Stabilised Ti Alloys

The second element in this case is more soluble in β-titanium than in α. The result is that the β-field is enlarged. The β-alloys can be further divided in two subgroups: isomorphous and eutectoid. In isomorphous systems, the second element and titanium are completely soluble in each other in the solid state, and hence the α to β phase transformation does not occur. The β-stabilizers are all BCC metals (vanadium, molybdenum, tantalum niobium) as may be expected. In eutectoid systems, there is partial solubility of the second element in β-titanium. The main alloying elements in this group are silicon and copper (which are termed fast acting) and chromium, manganese, iron, and nickel (all slow acting). The phase diagrams resulting from alloying of titanium with β-stabilizers are shown in figure 8.6.

## 8.3  CLASSIFICATION OF TI-ALLOYS

Titanium alloys are classified into five groups namely: commercially pure titanium, α-alloys, near α-alloys, α–β-alloys, and β-alloys.

### 8.3.1  Commercially Pure Titanium

These contain a minimum of 99 % titanium with traces of carbon, hydrogen, nitrogen and oxygen. They have excellent corrosion resistance but only moderate strength (300 to 1000 MPa). This strength increases with the oxygen content. Their main use is as containers for acids.

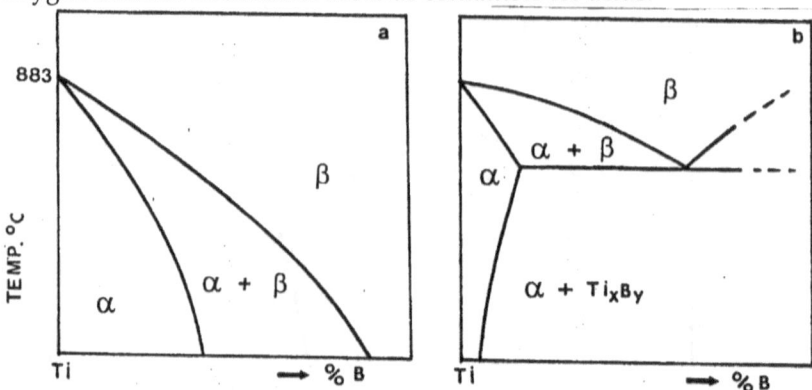

**Fig. 8.6** β-stabilized Ti-alloy systems: (a) isomorphous (b) eutectoid

### 8.3.2  α and Near α Alloys

Alpha alloys are mainly alloys of titanium with aluminium and or tin. Both of these are α-stabilizers and hence consist of the α-phase at all temperatures of commercial interest. As there is no phase change, the alloys are not heat treatable but gain their strength through solid solution strengthening. They have good weldability and oxidation resistance. The most common alpha alloy is Ti-5 % Al-2.5 % Sn.

Near alpha alloys are based on the isomorphous β-stabilized alloys. Up to 2 % of vanadium and/or molybdenum are added in addition to aluminium. The microstructure is mainly α but a small amount of β is retained down to room temperature due to the presence of the β-stabilizers. They are mainly used for aircraft manufacture (air frame and some engine parts) due to their high strength to weight ratio and their stability at elevated temperatures. They also have good weldability and toughness. Examples are Ti-8 % Al-1 % V-1 % Mo, and Ti-6 % Al-2 % Sn-4 % Zr-2 % Mo.

### 8.3.3 α-β Alloys

Here, sufficient amounts of both α and β stabilizers are added such that the microstructure consists of a mixture of the two. They may be heat treated by controlling the α to β transformation, and may also be welded, machined or forged. The most common α–β Ti-alloy is Ti-6 % Al-4 % V, which is also the most widely used titanium alloy. Others are Ti-6 % Al-6 % V-2 % Sn and Ti-6% Al- 2 % Sn-4 % Zr-6 % Mo.

The microstructure depends on the composition, method of processing and the heat treatment. This can be illustrated by considering the simplest of the alloys i.e., the Ti-6 % Al-4 % V alloy whose pseudo-binary phase diagram is shown in figure 8.7.

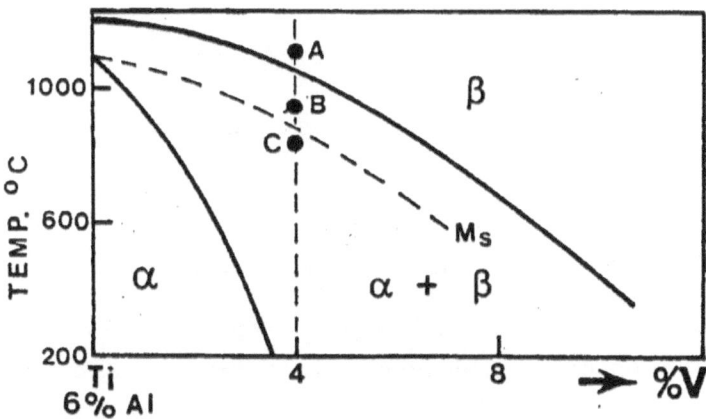

**Fig. 8.7** Heat treatment of α-β Ti alloys

Suppose the alloy is solution treated at about 1100 ºC and quenched. The β changes to α′ (called titanium martensite) by a shear mechanism. The resulting HCP structure appears as heavily twinned platelets. If on the other hand, the alloy is furnace cooled after solution treatment, the room temperature microstructure shall consist of an α matrix with intergranular (i.e., grain boundary) β phase.

If the solution heat treatment is done in the α–β phase field but above the martensite start ($M_s$) temperature (point B in figure 8.6) and then quenched, the microstructure will consist of α plus α′. The primary α is unaffected while the β transforms to α′ on quenching.

If solution heat treatment is done below the $M_s$ temperature (point C in figure 8.6), no changes take place and the final microstructure consists of well distributed α plus β.

The best mixture of mechanical properties (i.e., adequate strength combined with adequate toughness) is obtained by quenching from point B in figure 8.6. α–β titanium alloys are used mainly in aircraft manufacture.

### 8.3.4 Beta Ti-Alloys

When sufficient amounts of β-stabilizers (mainly chromium, vanadium, molybdenum or iron) are added, the β structure is maintained down to room temperature. These alloys are non-heat treatable but may be strengthened somewhat by cold working. They have low ductility and hence are not used much. Their main use is in the manufacture of high strength fasteners for the aircraft industry. The most widely used is Ti-13%V-11%Cr-3% Al.

## PROBLEMS

8.1 Describe the effects of alloying on the electrical conductivity and mechanical strength of copper.

8.2 With reference to the relevant part of the Cu-Zn equilibrium diagram, explain what α and α–β brasses mean. What are the major properties and applications of each?

8.3 Give an account of the 4 main bronzes containing elements other than tin, giving their main properties and uses.

8.4 Discuss with reference to the relevant phase diagram, the structure and uses of brasses for zinc contents varying from 10 to 45 %.

8.5 Describe the effects of alloying elements on the electrical conductivity, machinability, formability and mechanical strength of copper.

8.6 What alloying elements should be added to alloys of copper and zinc to improve their: (i) corrosion resistance (ii) strength (iii) machinability?

8.7 Knowing that aluminium and copper form two series of commercially useful alloys: one series being copper rich alloys of aluminium and the other, aluminium rich alloys of copper. Give a concise account of each group of alloys.

8.8 Describe the effects of alloying on the following properties with respect to pure copper: (i) electrical conductivity (ii) machinability (iii) formability (iv) mechanical strength (v) corrosion resistance.

8.9 What are some of the advantages and disadvantages of titanium as a base for engineering alloys?

8.10 Give an account of the various heat treatment procedures for $\alpha$-$\beta$ titanium alloys citing the microstructures that result from each procedure. Use the Ti-6%Al-4%V as an example and refer to the relevant portion of the phase diagram.

8.11 Explain what is meant by $\alpha$-Ti, near $\alpha$-Ti, $\alpha$-$\beta$ Ti and $\beta$-Ti alloys briefly mentioning the major applications of each.

## Further Reading

Higgins, R.A. Engineering Metallurgy Part I. Hodder and Stroughton, London, 1980.

Rollason, E.C. Metallurgy for Engineers. 4th. Ed., Edward Arnold, Norwich, 1973.

Smith, W. F. Structure and Properties of Engineering Alloys, 2nd ed. McGraw-Hill, New York, 1993.

# CHAPTER NINE

# HEAT TREATMENT OF STEELS

## 9.1  INTRODUCTION

As already stated, the term heat treatment refers to a series of heating and cooling operations which are performed on a material to achieve a given set of properties. Steels enjoy widespread use due to the multitude of properties they can have when given suitable heat treatment combined with appropriate alloying. It may be worthwhile at this point for the reader to revise chapters five and six.

## 9.2  TRANSFORMATION PROCESSES

In steels, the property changes during heat treatment are based on the decomposition of austenite. This decomposition is controlled by temperature and time dependent diffusion of iron and carbon atoms. Usually, equilibrium (i.e., formation of the products expected from the equilibrium diagram) is not reached. The products formed depend on the temperature at which decomposition takes place and the time available (or cooling rate) for the decomposition as follows:

### 9.2.1  Formation of Pearlite

When the decomposition of austenite takes place at the eutectoid temperature and the cooling rate is very low, simultaneous formation of α and cementite takes place. This takes place in the following steps (figure 9.1):

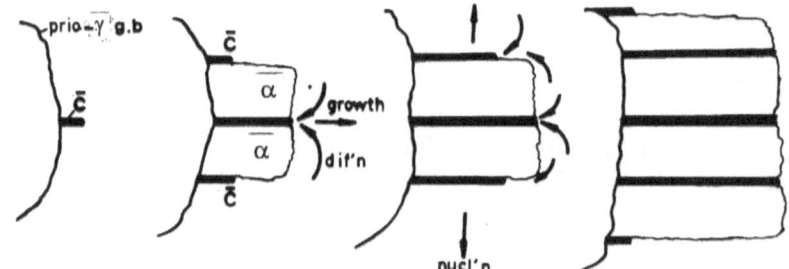

**Fig. 9.1** A schematic representation of the formation of pearlite

1. Nucleation of cementite starts at the prior austenite grain boundaries. This occurs because grain boundary atoms have more energy.

2. Carbon diffuses from the neighbouring grain into the cementite nucleus leaving these areas with less carbon. This leads to the formation of ferrite in these regions.

3. The nucleus grows longer by absorbing carbon from the areas ahead of it.

4. A new nucleus of cementite starts adjacent to the ferrite and grows in a similar manner.

As a consequence, the resulting microstructure consists of intimately mixed alternate layers of ferrite and cementite. This product, as has been stated already, is termed pearlite. It should be clear that the term pearlite refers only to the specific laminated mixture of ferrite and cementite formed from the decomposition of austenite of eutectoid composition at the eutectoid temperature. There are other mixtures of ferrite and cementite but since these are not lamellar, they are not termed pearlite.

The distance between the pearlite lamellae decreases (with a corresponding increase in hardness of the resulting product) when the cooling rate is increased. Fine pearlite is sometimes termed sorbite and even finer pearlite, troostite. Pearlite is a relatively soft phase.

## 9.2.2 Formation of Bainite

When the cooling rate is a little too high for the formation of the lamellae, the iron changes its crystal structure from FCC to BCC. Excess carbon latter diffuses to form dispersed particles of cementite inside a matrix of ferrite. The resulting microstructure is termed bainite and has adequate toughness (which results from the ability of the soft ferrite to deform) and adequate strength (from the ability of the dispersed cementite particles to block dislocation movement).

## 9.2.3 Formation of Martensite

If the cooling rate is above a certain critical value, and if the transformation takes place below a characteristic temperature (which is a function of the carbon content for plain carbon steels), the iron changes from FCC to BCC crystal structure by a shearing action. All the atoms move in unison by a few tenths of a nanometer. There is no diffusion involved hence the change is almost instantaneous. However, the solubility of carbon in FCC iron is about 100 times that in BCC iron. The excess carbon has no time to diffuse out of solution and form a new phase. The end result is that the carbon becomes trapped inside the BCC structure resulting in a supersaturated solution of carbon in BCC iron. The resulting phase, which has a body centred tetragonal crystal structure, is termed martensite. The critical cooling rate required for martensite formation is achieved by quenching the steel in water or oil from the austenite region.

Martensite is a non-equilibrium phase and hence is not shown anywhere in the equilibrium diagram. With time, it is expected to break down to the equilibrium phases ferrite and cementite. The rate of this transformation is however, negligible at room temperature but can be accelerated by heating in a process termed tempering.

The BCC unit cell is distorted by the presence of excess carbon making slip virtually impossible. Hence martensite is extremely hard and brittle. The hardness of martensite changes with increase in carbon content as shown in figure 9.2. Up to 0.8 % C (eutectoid composition), there is a monotonic increase. Above this composition, the hardness may either remain constant or decrease with increase in carbon content. If the steel is quenched from the austenite area, the hardness decreases due to the increasing quantities of retained austenite (a soft phase) in the steel. If quenching is from the two phase (γ plus cementite) area, the hardness remains constant. The austenite changes to martensite on quenching while cementite is retained. Since the hardness of cementite is approximately equal to that of martensite, the hardness remains the same.

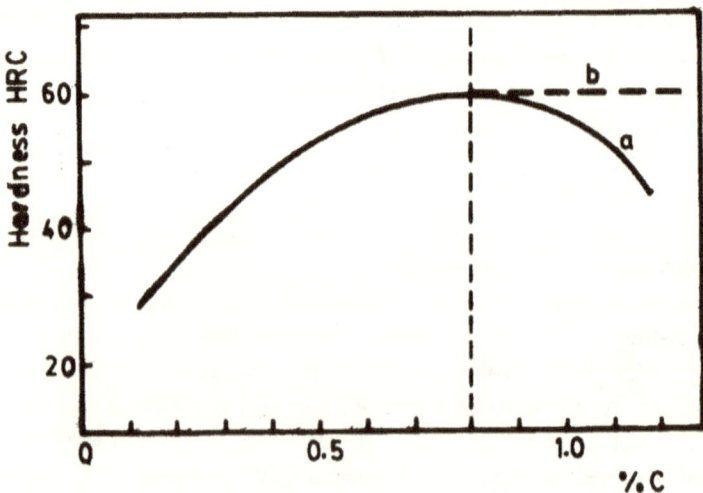

**Fig. 9.2** Variation of hardness with carbon content in steels: (a) Quenched from austenite region (b) Quenched from austenite + cementite region.

## 9.3  ISOTHERMAL TRANSFORMATION AND TTT CURVES

The formation of martensite described above is only complete if a certain cooling rate is achieved. This critical cooling rate depends on the carbon content of the steel as shown in figure 9.3. If the critical cooling rate is not achieved, transformation to martensite is only partial. Other products (ferrite, pearlite, bainite) are formed in addition to martensite. The formation of these other products involves breaking of existing bonds, diffusion of atoms to new sites and formation of new boundaries. This process of nucleation and growth requires time and energy, and since energy is a function of the temperature, the products of the transformation depend on the temperature at which transformation takes place as well as the time available for such transformation.

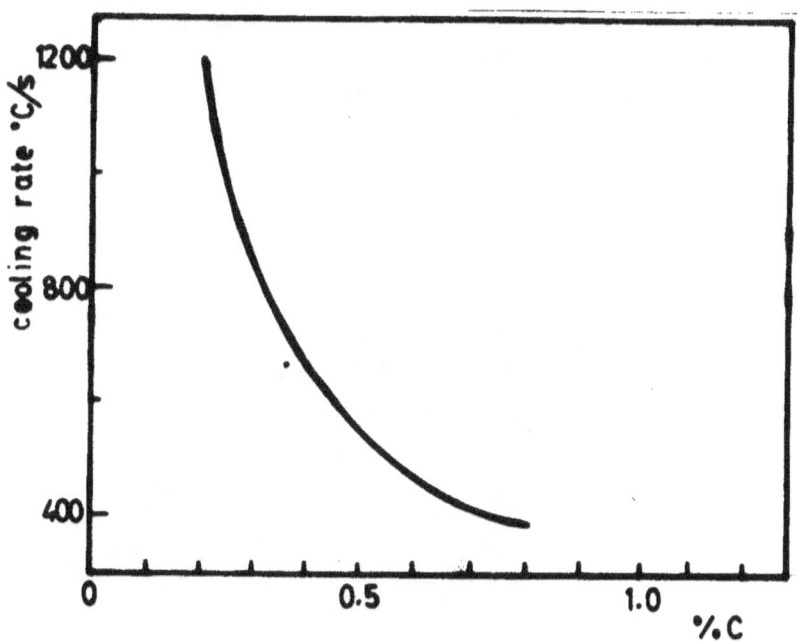

**Fig. 9.3** Variation of critical cooling rate with carbon content

Investigations to determine the nature of the products that result at various temperatures and the time required to initiate and complete the reactions are termed isothermal investigations and proceed as follows:

(i) A penny shaped piece of the material being investigated (say eutectoid plain carbon steel) is heated to austenizing temperature for a time long enough to become fully austenite (the penny shape helps to achieve uniform transformation).

(ii) The piece is quenched in a suitable bath (molten metal, salt, etc.) to the temperature being investigated (say 700°C).

(iii) It is kept at this temperature for a given time. If this time is sufficient, transformation to other products (say pearlite) will start.

(iv) The piece is then quenched in water. Any austenite that had not transformed in step (iii) above is transformed to martensite.

(v) The piece is observed under the microscope. The portion of the piece that is martensitic, x % can then be estimated. The extent of transformation in step (iii) is then (100 - x) %. Furthermore, from microscopic examination, the products of the transformation in step (iii) are determined.

(vi) Steps (i) to (v) are repeated while varying the time in step (iii). This will help determine the time required for transformation to start and the time required for transformation to be completed at the specific temperature chosen in step (ii).

(vii) Steps (i) to (vi) are repeated for different temperatures.

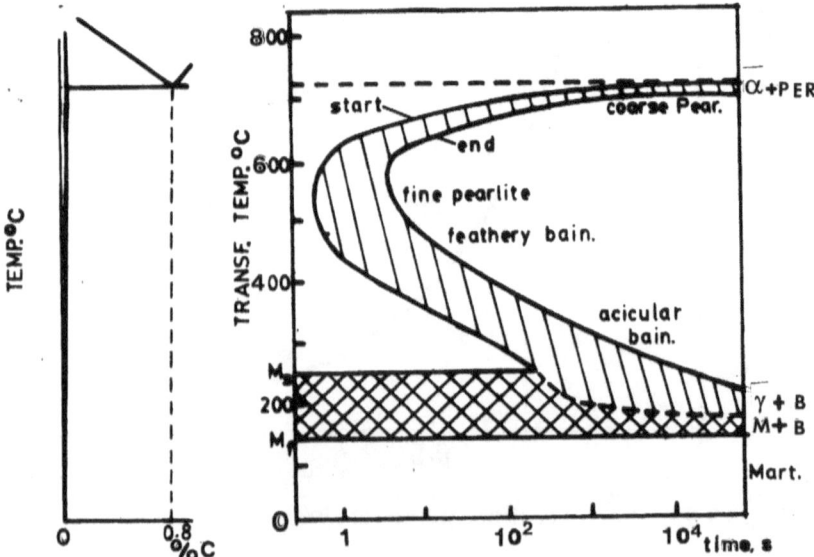

**Fig. 9.4** TTT curve for eutectoid steels.

The results of isothermal investigations are plotted on isothermal transformation diagrams, which are also called TTT (Temperature Time Transformation) curves. For eutectoid plain carbon steel, this has the form shown schematically in figure 9.4. It is noted that the time required for transformation to start (and end) decreases with temperature up to a certain value, and then increases again. This gives the diagram a C-shape and hence TTT curves are also termed C-curves. The point on the TTT curve nearest the temperature axis is termed the knee or nose of the curve. It is also noted that martensite can only form in steel when the transformation takes place below a certain temperature, $M_s$, and that 100 % martensite will only result if the transformation takes place below another characteristic temperature, $M_f$. Both $M_s$ and $M_f$ are functions of the carbon content in plain carbon steels.

The TTT curves for hypo- and hypereutectoid steels have the same shape as above but with the additional formation of ferrite or cementite as shown in figure 9.5 for hypo eutectoid steel. TTT curves are useful in deciding heat treatment procedures in steel manufacture.

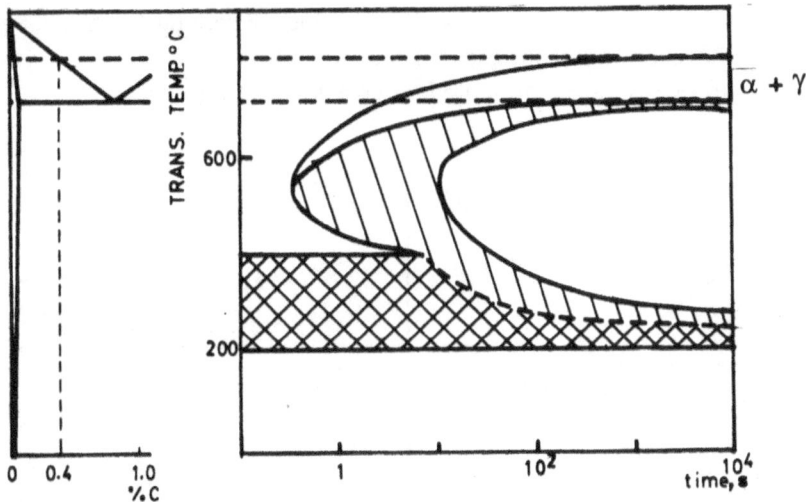

**Fig. 9.5** TTT curve for hypereutectoid steel

## 9.4 TECHNICAL HEAT TREATMENT PROCEDURES FOR STEELS

The following heat treatment procedures are applied commercially to steels:

### 9.4.1 Annealing

This involves heating the steel to a temperature where diffusion is easier and the atoms can rearrange themselves into more stable structures.

Annealing is followed by very slow cooling (in the furnace). There are several specific types of annealing:

Full annealing: The material is heated to austenizing temperature followed by furnace cooling. The final structure is coarse pearlite. The procedure is used to soften prior to machining.

Process annealing: Involves heating to a temperature just below the lower critical temperature for hypoeutectoid steels and a temperature oscillating on the lower critical temperature for hypereutectoid steels. Secondary and eutectoid cementite are transferred to a more stable globular form. Process annealing is done to improve machinability and cold working.

Stress-relief annealing: The material is heated to a temperature below 650 °C followed by furnace cooling. The purpose is to reduce internal (residual) stresses in a machine part without changing the internal structure. The process follows all manufacturing processes where the work piece is subjected to non-homogenous heating or cooling treatments e.g., welding, hot working, casting (steel), etc.

These processes are represented schematically in figure 9.6.

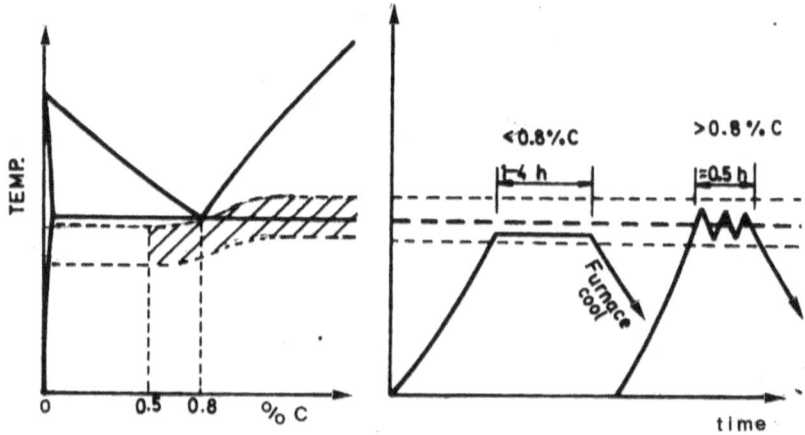

**Fig. 9.6** Time-temperature curve for process annealing

## 9.4.2 Normalizing

The heating range for normalizing is the same as for full annealing but is followed by air-cooling. Due to the faster rate of cooling, a finer grain structure is obtained. The mechanical properties, especially the fatigue strength, are improved. Normalizing reverses all structural changes introduced during various manufacturing processes such as hot and cold working, welding, hardening or tempering. The process is shown schematically in figure 9.7.

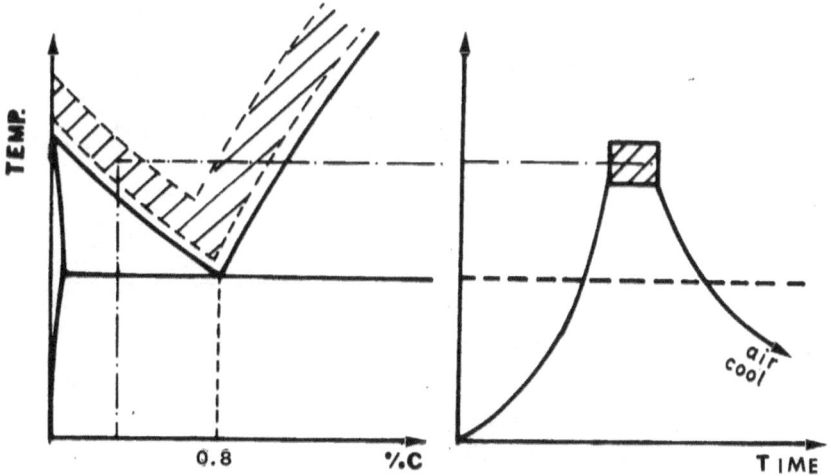

**Fig. 9.7** Time-temperature curve for normalising

### 9.4.3 Quench Hardening

This involves cooling of the steel from a temperature above G-S for (hypo-eutectoid steels) or above S-K (for hyper-eutectoid steels) with a cooling rates fast enough in order to form martensite. The quenching media is either water or oil and the cooling rate has to exceed the upper critical cooling rate (see figure 9.8). Direct quenching often leads to quench cracks which form as follows: on quenching, the cooling rate is highest on the surface. Martensite will therefore form on the surface while the core is still austenite. Later, martensite forms at the core, and since the density of austenite is higher than that of martensite, this change will be accompanied by an expansion of the core. This will induce stress on the brittle martensite already formed on the surface leading to quench cracks.

In plain carbon steels with diameters in excess of 10 mm, the core may not transform to martensite. Alloying additions can increase the thickness of the surface layer transformed to martensite, as we will see later. Alloying additions also reduce the critical cooling rate required for formation of martensite hence avoids quench cracks. Other methods of avoiding quench cracks are martempering and austempering considered in the following sections.

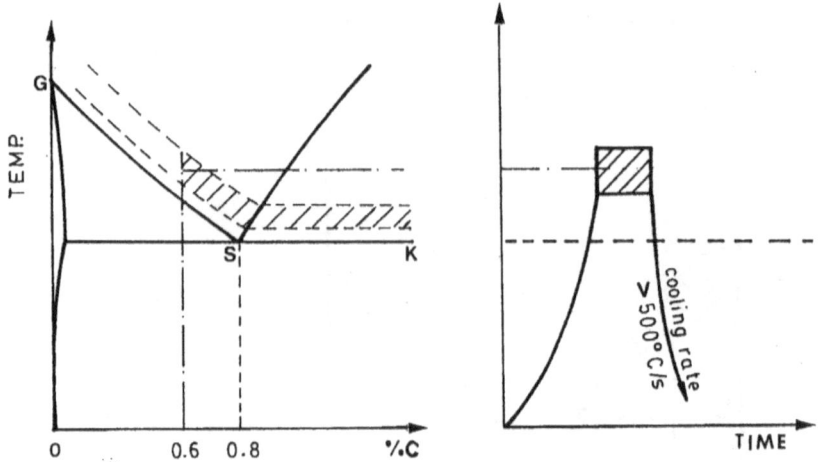

**Fig. 9.8** Time-temperature curve for quenching

### 9.4.4 Martempering (Interrupted Quenching)

The steel is quenched in a salt bath to a temperature above $M_s$ thereby avoiding the knee of the TTT curve. It is then cooled at a much slower rate resulting in a more uniform cooling and hence avoiding quench cracks (see figure 9.9).

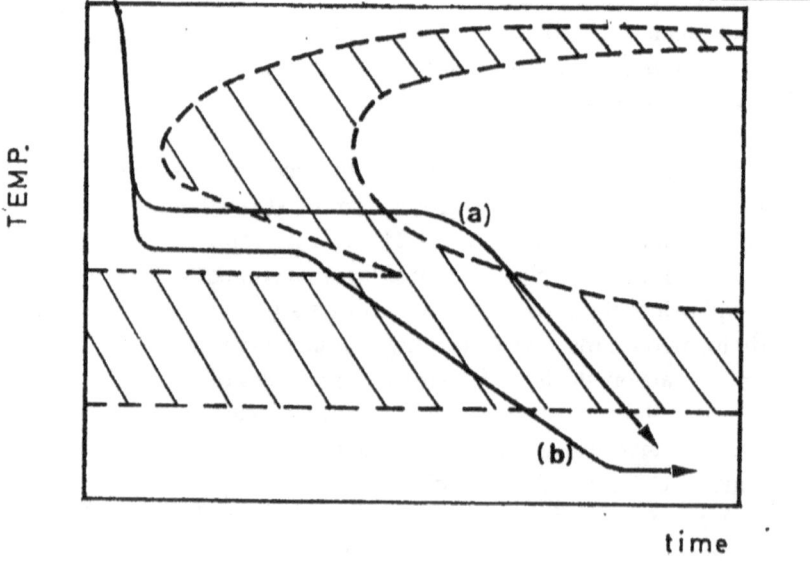

**Fig. 9.9** (a) Austempering and (b) martempering

## 9.4.5 Austempering (Isothermal Tempering)

As in martempering, the steel is quenched to above $M_s$ (250 to 350 °C). It is then held at this temperature until transformation to bainite is complete as shown if figure 9.9. Ferrite and cementite then form by a mixture of shear and diffusion with the cementite forming dispersed particles inside ferrite. The resulting structure is similar to tempered martensite and hence austempering combines quenching and tempering in one operation.

## 9.4.6 Tempering

Martensite is too brittle for most applications. To impact some toughness on it, it is heated to a temperature below the eutectoid temperature (typically 200 °C to 700 °C), followed by air cooling or quenching (alloy steels). Internal stresses are relieved and any retained austenite decomposes. The unstable martensite regains the more stable cubic lattice structure by precipitating cementite. The resulting structure is ferrite with fine dispersed spheroidal carbide particles. The increase in toughness is accompanied by a decrease in hardness.

Proper control of tempering temperature and time are required to prevent temper brittleness that results from formation of carbides (Mn, Ni & Cr) at grain boundaries. The specific tempering temperature and time depends on the composition of the steel and the desired properties. The higher the tempering temperature the greater the loss of hardness. This is accompanied by an increase in the size of cementite particles. The time-temperature curve for tempering is shown in figure 9.10.

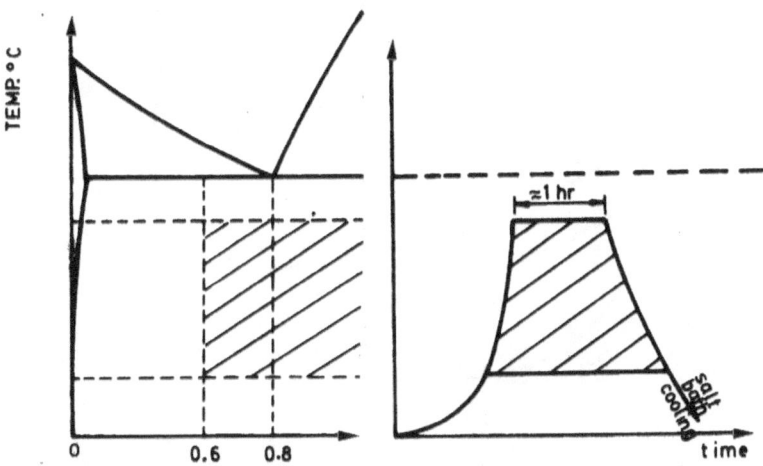

**Fig. 9.10** Time-temperature curve for tempering

### 9.4.7   Surface Hardening

This refers to a procedure in which only the surface is hardened leaving the core soft. The advantages are that a tough core is combined with a hard surface (which increases the resistance of the surface to wear and contact pressure), the danger of quench cracks is decreased, and the fatigue strength is increased by introducing internal compressive stresses into the surface layer. The process is also economical since only partial hardening is undertaken. The process is useful for components like gears, crankshafts, cylinder liners, dies, etc.

Surface hardening can be accomplished by flame hardening, induction hardening, case hardening, or nitriding. In flame and induction hardening, the surface of the work is heated either by gas burner (flame) or by electrical induction followed by quenching. The method is suitable for unalloyed steels with 0.35 % to 0.6 % C. In the other processes, the chemical composition of the surface is changed. Carburizing involves heating a low carbon, (< 0.2 % C) unalloyed or low-alloyed steel to austenizing temperature in a carburizing medium. The carbon diffuses into the metal and on quenching martensite forms on the surface. The carburizing agents include: a mixture of charcoal and $BaCO_3$, $Na_2CO_3$ or $MgCO_3$, a salt bath containing NaCN plus $BaCl_2$, gaseous hydrocarbons ($CH_4$, $C_2H_6$) plus nickel catalyst. In nitriding, the component is heated to about 550°C in either gaseous ammonia or a cyanate containing salt bath. Steels for nitriding contain aluminium (about 1 %) or chromium (1 to 3 %) and Molybdenum (0.2 to 1 %), which, form nitride precipitates. A very hard surface (> 1100 HV) results, and no quenching is necessary.

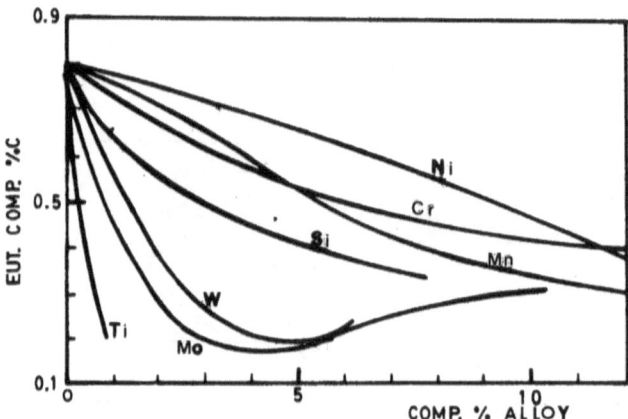

**Fig. 9.11** The effect of alloying elements on the eutectoid carbon content

## 9.5  EFFECTS OF ALLOYING ELEMENTS.

The main purpose of adding alloying elements to steel is to enable martensite to form at a lower cooling rate, thus improving the hardenability of the steel and avoiding quench cracks. Steels that have only carbon (and up to 1 % Mn) as alloying element are termed a plain carbon steels. Those with up to 5 % other alloying elements are low alloy steels while high alloy steels have more than 5 % other alloying elements. The alloying elements added to steel include Cr, Mn, Ni, Mo, Si, V, W, Co, Ti.

All alloying elements shift the TTT-curves to the right by inhibiting diffusion of carbon, lowering the $M_s$ Temperature, and increasing the tendency of the pearlite and bainite knees to separate. They also lower the eutectoid carbon content (figure 9.11). They may either lower or raise the eutectoid temperature depending on their crystal structure. Ti, Mo, Si, W and Cr raise the eutectoid temperature. All are soluble by substitution in ferrite and hence extend the range over which ferrite is stable. They are termed α-stabilizers. Ni, Co and Mn on the other hand are soluble by substitution in γ and lower the eutectoid temperature. Since they extend the range over which γ can exist, they are termed γ-stabilizers (figure 9.12).

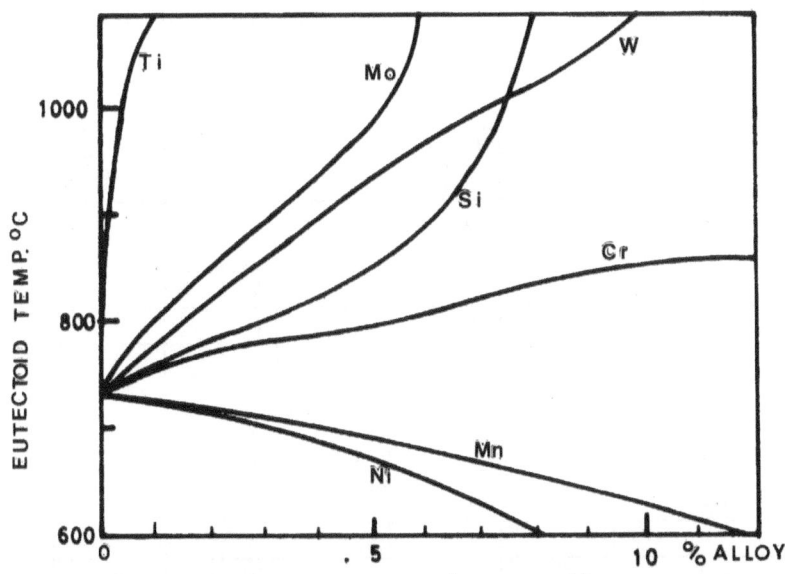

**Fig. 9.12** Effect of alloying elements on the eutectoid temperature

Most alloying elements (excluding Ni, Al and Si) form carbides that are harder than cementite and hence increase the strength of steel without

**Table 9.1** The effect of alloying elements on properties of steel

| Ele | Har | YS | TS | δ | ψ | $C_v$ | CR |
|---|---|---|---|---|---|---|---|
| Si | 1 | 2 | 2 | -1 | C | -1 | -1 |
| Mn (PS) | 1 | 1 | 1 | C | C | C | -1 |
| Mn (AS) | -3 | -1 | 1 | 3 | C | -- | -2 |
| Cr | 2 | 2 | 2 | -1 | -1 | -1 | -3 |
| Ni (PS) | 1 | 1 | 1 | C | C | C | -2 |
| Ni (AS) | -2 | -1 | 1 | 3 | 2 | 3 | -2 |
| Al | -- | -- | -- | -- | -1 | -1 | -- |
| W | 1 | 1 | 1 | -1 | -1 | C | -2 |
| V | 1 | 1 | 1 | C | C | 1 | -2 |
| Co | 1 | 1 | 1 | -1 | -1 | -1 | 2 |
| Mo | 1 | 1 | 1 | -1 | -1 | 1 | -2 |
| Cu | 1 | 2 | 1 | C | C | C | -- |
| S | -- | -- | -- | -1 | -1 | -1 | -- |
| P | 1 | 1 | 1 | -1 | -1 | -3 | -- |

| Ele | CF | WR | FG | MC | NIT | Cor |
|---|---|---|---|---|---|---|
| Si | -1 | -3 | -1 | -1 | -1 | -- |
| Mn (PS) | C | -2 | 1 | -1 | C | -- |
| Mn (AS) | -- | -- | -3 | -3 | -- | -- |
| Cr | 2 | 1 | -1 | -- | 2 | 3 |
| Ni (PS) | -- | -2 | -1 | -1 | -- | -- |
| Ni (AS) | -- | -- | -3 | -3 | -- | 2 |
| Al | -- | -- | -2 | -- | 3 | -- |
| W | 2 | 3 | -2 | -2 | 1 | -- |
| V | 4 | 2 | 1 | -- | 1 | 1 |
| Co | -- | 3 | -1 | C | -- | -- |
| Mo | 3 | 2 | -1 | -1 | 2 | -- |
| Cu | -- | -- | -3 | C | -- | 1 |
| S | -- | -- | -3 | 3 | -- | -1 |
| P | -- | -- | -1 | 2 | -- | -- |

Notes on table 9.1:

| | |
|---|---|
| Ele = Alloying element | ψ = Reduction in area |
| YS = Yield stress | TS = Ultimate tensile strength |
| δ = Percent elongation | $C_v$ = Impact energy |
| CR = Critical cooling rate | CF = Carbide formation |
| WR = Wear resistance | FG = Forgeability |
| MC = Machinability | NIT = Nitrability |
| Cor = Corrosion resistance | PS = Pearlitic steel |
| AS = Austenitic steel | -- = Unknown/uncharacteristic |
| Negative = Reduction | C = constant/ unaffected |
| Higher number = Greater effect | |

lowering the ductility. Ni, Al and Si promote formation of graphite and are added only to steels with very low carbon or when appreciable quantities of carbide formers are also present. The effects of alloying elements on the mechanical and other properties are summarized in Table 9.1.

## 9.5.1 Stainless Steels

These are a special class of alloy steels with the main property being their resistance to corrosion. This property makes them useful in the manufacture of sinks, cutlery, food containers, etc. They contain at least 12 % Cr which, on contact with air, forms a thin impervious layer of CrO. This protects the steel from further attack and gives the article a high polish appearance.

Adding nickel increases the corrosion resistance. Since chromium is an α-stabilizer while nickel is a γ-stabilizer, the resulting microstructure depends on the proportions of these two elements in the steel as shown in figure 9.13. Generally, three main groups of stainless steels can be distinguished:

Ferritic stainless steels (or chromium steels). Chromium predominates and hence the mirostructure stays BCC at all temperatures. These steels cannot be hardened by heat treatment since no α --> γ transformation can take place. They cannot be grain refined either. The carbon content is kept low (about 0.08 %) to prevent carbide precipitation.

Austenitic stainless steels (or Ni--Cr steels). Contain appreciable amounts of both nickel and chromium (minimum 8 % Ni; minimum Ni + Cr = 24 %). The microstructure stays FCC down to room temperature and again heat treatment is not possible. They have better corrosion resistance; improved workability compared to ferritic steels and are weldable. Being austenitic at room temperature, they are non-magnetic.

Martensitic stainless steels. These contain the proportions of nickel and chromium shown in figure 9.13. Under these conditions, the α to γ transformation is possible and hence heat treatment can be carried out. The carbon content has to be above 0.15 %.

Other α-stabilizers have the same effect as chromium and similarly, other γ- stabilizers have the same effect as nickel. To take these into account, the chromium and nickel equivalents are defined. These are calculated from: $Ni_{eq} = \% \, Ni + 0.5 \times \% \, Mn + 30 \times (\% \, C + \% \, N)$ and $Cr_{eq} = \% \, Cr + \% \, Mo + 1.55 \times \% \, Si + 0.5 \times \% Nb$. It is these equivalent values that are shown in figure 9.13.

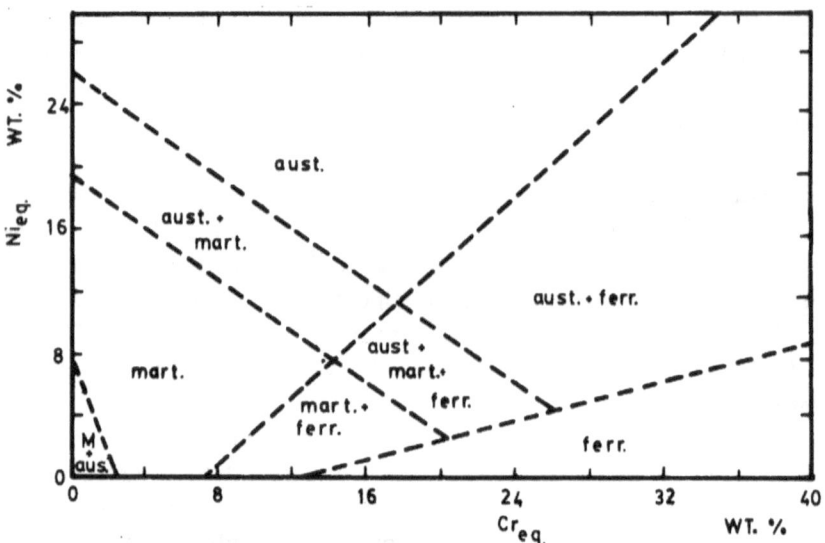

**Fig. 9.13** Change of microstructure of stainless steels with nickel and chromium contents

**Weld decay:** When stainless steels are welded, the heat-affected zone (HAZ) is heated to the range 500 °C to 700 °C. In this temperature range, chromium precipitates carbides at the grain boundaries (especially in steels with appreciable carbon content). The region near the grain boundary is left without chromium and hence liable to corrosion. Thus welded stainless steel corrodes in the HAZ. This phenomenon is termed weld decay. To prevent this, the article can be reheated to about 1000 °C after welding (to re-dissolve the chromium) and then quenched. Alternatively, a stabilized stainless steel can be used (i.e., a steel to which titanium or niobium are added). These are better carbide formers than chromium and form carbides preferentially, leaving chromium in solution.

## 9.6 CLASSIFICATION OF ALLOY STEELS.

There are many criteria that are used as the basis for classification of steels. Some of these are: Chemical composition (e.g., nickel or manganese steels, etc.); Properties (e.g., stainless, high strength, etc.); Structure (pearlitic, martensitic, ferritic, etc.); Uses (tool steels, spring steels, structural steels, etc.).

Various standardization bodies have developed classification codes for steels. For example, in the German Standards (DIN), non-alloyed steels are designated by their minimum tensile strength. For example, St. 37 is a structural plain carbon steel with a minimum tensile strength of 370 MPa. Low alloy steels are designated by their chemical composition. Multiplying factors are implied for different elements. These are: 4 for Cr, Co, Mn, Ni, Si, W; 10 for Mo, Ti, V; and 100 for C). As an example, 34 Cr 4 = steel with $^{34}/_{100} = 0.34$ % C and $^4/_4 = 1$ % Cr; 18 Cr Ni 8 = steel with 0.18 % C, 2 % Cr and 2 % Ni. High alloy steels are identified by a preceding letter X and the percentages are not multiplied by the factors e.g., X 12 CrNi 18 8 means a high alloy steel containing 0.12 % C, 18 % Cr and 8 % Ni.

In American standards (AISI: American Iron and Steel Institute or SAE: Society of Automobile Engineers) the code consists of 4 digits. The first two identify the alloy group while the last two indicate the carbon content multiplied by 100 e.g., AISI 1060 is a plain carbon steel (10) with 0.6 % C. AISI 4340 on the other hand is an alloy steel containing nickel and molybdenum (43) and 0.4 % carbon.

In British Standards (BS 970) a six-digit system is used. The first 3 digits represent the manganese content x 1000. Digit 4 is a letter: A if the steel is classified by chemical composition; M if specified by mechanical properties or H if specified by hardenability. Digits 5 and 6 represent carbon content multiplied by 100 e.g., 060 A 40.

At the time of writing no Kenya Standard on classification of steels had been written, but one is expected soon.

## 9.7 HARDENABILITY AND RULING SECTION.

When a component is quenched and then tempered, optimum properties are obtained only if there is 100 % conversion to martensite (after quenching). As has been stated already, formation of martensite requires a critical cooling rate (which depends on the carbon content and alloying additions). When a large piece is quenched, the critical cooling rate may be achieved on the surface but not at the centre. As a result, there may only be partial conversion to martensite at the centre. If a hardness profile is taken across the cross section, a progressive decrease in hardness is noted as the centre of the piece is approached.

For a material with a given cross section subjected to a defined quenching procedure, there is a maximum cross section that can become fully martensite on quenching. Any larger section will have only partial conversion to martensite at the centre. This effect is termed mass effect. The maximum diameter of a round bar that can be treated to a given set of mechanical properties is termed the limiting ruling section.

From the above, it is evident that we need to determine the hardenability of given steel. This is a measure of the ease with which the steel can be quenched to a fully martensite structure. Two methods are used to determine the hardenability: the Grossmann method and the Jominy end quench method. In the first method, the microstructure is observed in several quenched bars of various diameters to assess the extent of conversion to martensite. The Jominy end quench test is used more often. Here, a round bar with the dimensions shown in figure 9.14 is heated to austenizing temperature and then quenched from one end with a jet of water (British Standard BS 4437). The cooling rate is highest at the quenched end. The hardness is then determined along the bar and the results plotted in a hardenability curve such as that shown in figure 9.15. It may be noted from the figure that the alloy steel has a better hardenability than the carbon steel. The hardenability curves can be used to estimate the depth of hardness, which can be expected from a machine part made from the steel.

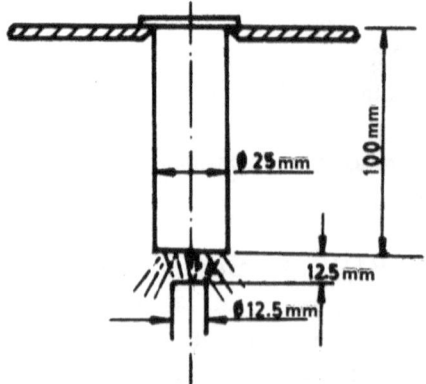

**Fig. 9.14** The Jominy end-quench test setup

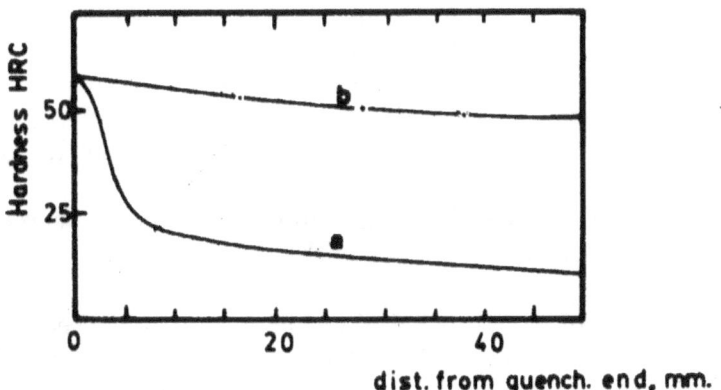

**Fig. 9.15** Hardenability curves (a) plain carbon steel (b) alloy steel

## PROBLEMS

9.1 Sketch the following heat treatment processes and indicate the effects of each processes on the grain structure and mechanical properties of steel: (i) process annealing (ii) tempering (iii) quench hardening.

9.2 Distinguish between hardness and hardenability. Explain the procedure and interpret the results of the Jominy end quench test on 0.5 % C steel. Describe and explain the microstructure expected along the bar.

9.3 (a) Give an account of how isothermal transformation diagrams for steel may be compiled.
(b) Explain how quench cracks develop and describe in detail three methods for avoiding such cracks.

9.4 (a) Give an account of the formation of Martensite in steels.
(b) Sketch a TTT curve for eutectoid steel showing all likely products of the reactions.

9.5 Explain what is meant by the term weld decay in stainless steels and give two methods that can be used to prevent it.

9.6 Describe the crystal structure that would result after each stage of the following treatment when performed successively on a penny shaped piece of 0.4 % C steel:

i) Heat to 830 °C and hold at this temperature for 30 minutes;

ii) Quench to 600 °C and hold for 5 seconds;

iii) Continue holding at 600 °C for 15 seconds more;

iv) Quench in water to room temperature;

v) Heat to 500 °C, hold there for 30 minutes then cool to room temperature;

vi) Heat to 830 °C; keep at this temperature for 30 minutes, then cool slowly in the furnace.

9.7 Give an account of the formation of pearlite in eutectoid steel cooled under equilibrium conditions. What is the main difference between the resulting structure and other combinations of ferrite and cementite?

9.8 Sketch and fully label the TTT curve for 0.4 % C steel. With reference to the said diagram, explain the processes of austempering and martempering. What are the advantages of these processes over direct quenching?

9.9 Why would you expect the hardness of hyper-eutectoid plain carbon steels to decrease with increase in carbon content when quenched from the fully austenite state? What can you do to prevent such decrease and why?

9.10 Sketch the following heat treatment processes and indicate the effects of each process on the grain structure and mechanical properties (strength, hardness, toughness and ductility) of plain carbon steels: (i) Normalizing. (ii) Quench hardening. (iii) Tempering.

9.11 Distinguish clearly between the following pairs of terms: (i) normalizing and air hardening (ii) austempering and martempering (iii) hardness and hardenability.

9.12 Give an account of how isothermal transformation diagrams for steel may be compiled. Draw and label such a diagram for 1.2 % C steel.

9.13 Explain the micro-structural changes you would expect to occur when ordinary stainless steel is welded. Explain the effects of these changes on the corrosion properties of the steel and how they could be avoided.

9.14 (a) Explain what is meant by martensite and why it is not shown anywhere on the Fe-C equilibrium diagrams.

(b) Sketch and label an isothermal transformation diagram for a 0.4 % C steel and explain how such a diagram can be constructed.

9.15 Explain what is meant by the terms: (i) austenitic stainless steels (ii) weld decay (iii) martempering.

## Further Reading

Pascoe, J.K. An Introduction to the Properties of Engineering Materials, 3rd ed. ELBS & Van Nostrand Reinhold (UK), Workingham, 1982.

Higgins, R.A. Engineering Metallurgy, Part I, 4th Ed. Hodder & Stoughton, London, 1980.

Callister, W.D. Materials Science and Engineering, 3rd ed. John Wiley and Sons, New York, 1994.

Schaffer, J. P. et al. The Science and Design of Engineering Materials, Irwin, Chicago, 1995.

# CHAPTER TEN

# FATIGUE AND FATIGUE FAILURES

## 10.1   INTRODUCTION

Most components of interest to mechanical engineers are in a state of motion during their service. Common examples include shafts, gears, axles, piston rods, etc. Such movement usually means that the stress in these components is constantly fluctuating. Under fluctuating stress, failure occurs at much lower stresses than those predicted from strength of materials theory. This phenomenon is termed fatigue, and is responsible for an estimated 90 % of all service failures.

The surface of a fatigue fracture has the characteristic appearance shown in figure 10.1. There is a spot where the crack initiated (this is usually a point of stress concentration like a key way, oil hole, tool mark, etc., on the surface); a smooth part with "beach marks" where the crack propagated while the component was in service; and a rough part showing the final fracture. The beach marks on the propagation surface result from differences in environmental attack on the exposed fracture surfaces. The final fracture (rough surface) takes place when the crack has grown to an extent that the net section area cannot support the load any longer.

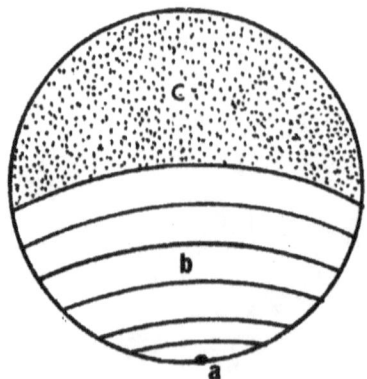

**Fig. 10.1** Fatigue fractured surface (a) point of crack initiation (b) crack propagation (c) final fracture.

The stress fluctuation in a component may vary differently with time as shown in figure 10.2. In a shaft rotating at constant speed, the variation is sinusoidal. The pulsating, repeated and random variation (such as would be experienced by the wing of an aircraft as it moves through the air or part of the undercarriage of a car as it hits randomly scattered pot-holes and bumps) cycles are as shown in the figure.

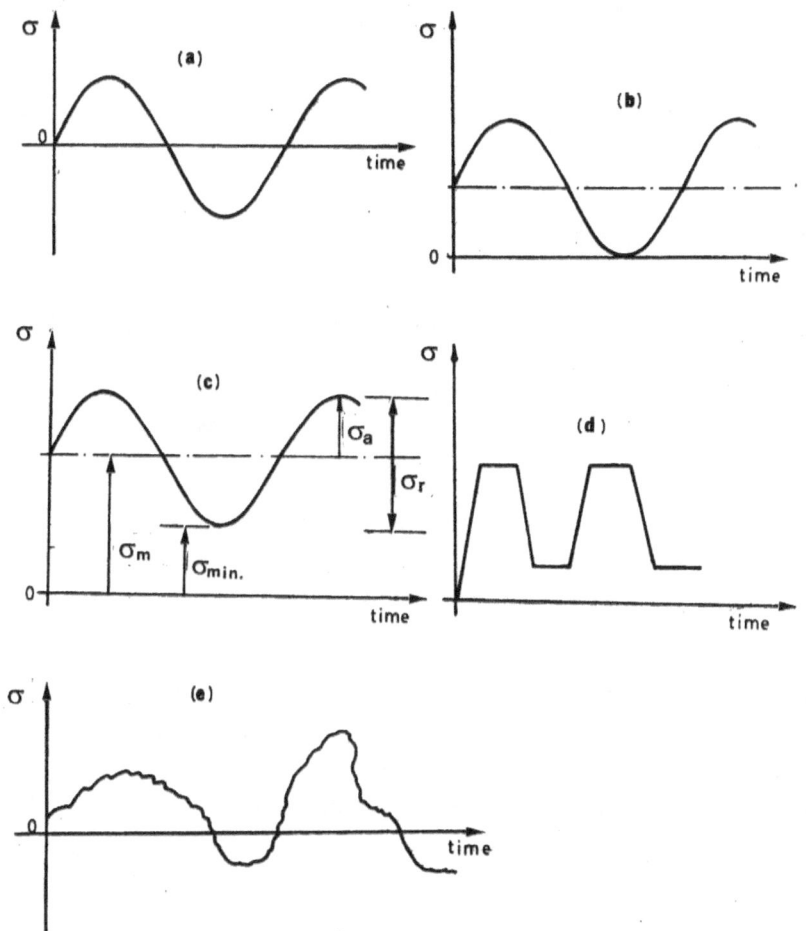

**Fig. 10.2** Various stress cycles in fatigue (a) completely reversed (b) pulsating (c) repeated (d) square wave form (e) random

The following variables may be defined:
(a) The maximum stress, $\sigma_{max}$, is the highest stress attained
(b) The minimum stress, $\sigma_{min}$, is the lowest stress attained
(c) A variation of stress from the minimum to the maximum and back to

the minimum is termed a stress cycle.

(d) The stress range, $\sigma_r$, is the difference between the maximum and minimum stresses.

(e) The alternating stress, $\sigma_a$, is half of the stress range, i.e:

$$\sigma_a = \frac{\sigma_r}{2} = \frac{\sigma_{max} - \sigma_{min}}{2} \qquad (10.1)$$

(f) The mean stress, $\sigma_m$, is the mean of the maximum and minimum stresses, i.e:

$$\sigma_m = \frac{\sigma_{max} + \sigma_{min}}{2} \qquad (10.2)$$

(g) The stress ratio, R, is the ratio minimum stress to the maximum stress, i.e:

$$R = \frac{\sigma_{min}}{\sigma_{max}} \qquad (10.3)$$

A fatigue-fractured surface always appears to be largely perpendicular to the maximum principal stress even in ductile materials. However, on a micro-scale, the mechanism is revealed to be by shear. Details of these mechanisms shall not be considered here.

## 10.2   LOW CYCLE AND HIGH CYCLE FATIGUE

At low values of alternating stress, the stress is proportional to the strain (elastic region). Thus either the stress or the strain could be used for fatigue analysis. A high number of stress cycles are required to cause failure and hence these circumstances are termed high cycle fatigue (HCF). When the stress is fairly high and considerable macroscopic plasticity is involved, the proportionality between stress and strain can no longer be assumed. Here, the strain range, rather than the stress range has to be used for fatigue analysis. In this case, failure occurs after a comparatively low number of stress cycles (typically less than $10^4$ cycles) and is termed low cycle fatigue (LCF). LCF analysis is useful for components like pressure vessels (both nuclear and ordinary), some components of aircraft e.g., landing gear, etc., which are expected to undergo only a limited number of stress reversals in their service life.

## 10.3   HCF TESTING (S-N CURVES)

To enable engineers to design against fatigue failure, fatigue data of materials have to be generated. This is normally done by testing small polished specimen of the form shown in figure 10.3 in rotating bending (hence the fatigue test is sometimes referred to as the rotating bending test). The specimen is loaded by a constant weight at one end and then rotated. Thus the specimen experiences a bending stress and due to the distribution of bending stresses part of the specimen is in tension while the other is in compression. On rotating, the stress in a particular fibre will change from the maximum tensile value to the maximum compressive value (depending on the fibers' position). The end result is a completely reversed sinusoidal stress variation. The shape of the specimen ensures the maximum stress occurs at the smallest portion.

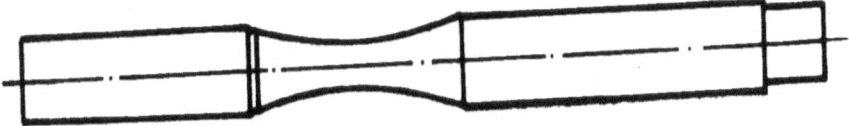

**Fig. 10.3** General shape of the fatigue specimen

The specimen is rotated until failure occurs and the stress recorded against the number of stress cycles required to cause failure, N. This is repeated for several specimens (at the same stress level) and for various stresses. The results are plotted on an S-N (stress-number of cycles) curve which has the typical shape shown in figure 10.4. (Note: the N-axis is usually on log scale.) The figures, also termed Woehler curves show that there is a general decrease in the number of cycles to failure as the alternating stress is increased and that for steels, the points fall roughly on a straight line. Furthermore, in steels, a level is reached below which no fracture occurs. This stress is termed the endurance limit, $\sigma_e$. This value may therefore be used in design against fatigue in the same manner as the yield stress is used in design against yielding.

Figure 10.4 shows that for aluminium alloys and other non-ferrous metals, no distinct endurance limit is observed. To design against fatigue in these alloys, fatigue strength at a given number of cycles, usually $10^6$ is determined and used in place of the endurance limit.

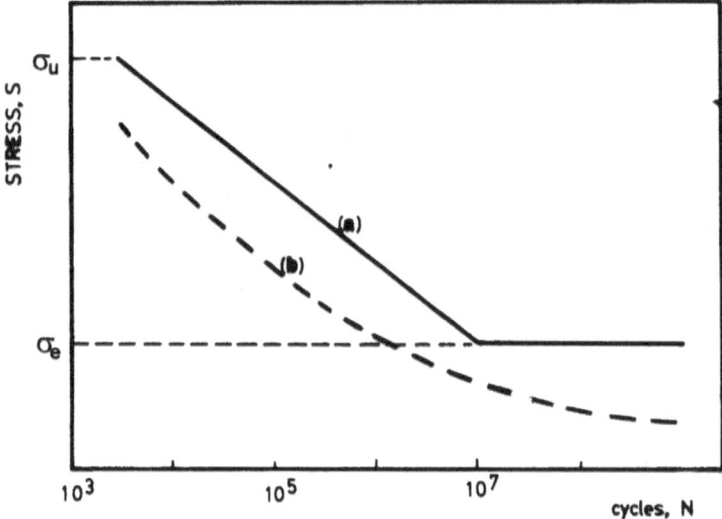

**Fig. 10.4** S-N curves: (a) ferrous metals (b) non-ferrous metals

There is considerable scatter in fatigue data. This mainly results from the dependence of fatigue crack initiation (nucleation) on the microscopic properties at the surface of the material. Thus in some analysis of fatigue data a statistical approach is used. In this approach, the S-N curve is drawn for a particular probability i.e., the line is drawn such that 90 % (say) of all points are above it. A component designed using this approach is deemed to have 90 % level of confidence against fatigue failure.

In general, the endurance limit (or fatigue strength at a given number of cycles) increases with increase in the material's ultimate tensile strength. As a rough guide, $\sigma_e = 0.5\sigma_{UTS}$ for steels if $\sigma_{UTS}$ is less than 1400 MPa and is equal to 700 MPa otherwise; $\sigma_e = 0.4\sigma_{UTS}$ for cast iron and $\sigma_e = 0.3$ to $0.4\sigma_{UTS}$ for aluminium alloys.

## 10.4   FACTORS AFFECTING ENDURANCE LIMIT
The endurance limit described above is affected by many factors. The main ones are the type of loading, the size of the component, the surface finish and the mean stress.

### 10.4.1 Type of Loading
Taking the case of bending as reference, the endurance limit is observed to be less when another type of loading e.g., fluctuating tension is applied. This reduction in endurance limit can be attributed to the following factors:

(i) In tensile loading, the whole cross section is subjected to the total stress i.e., the stress distribution is the same throughout the cross-section. This contrasts with bending loading where only the outermost fibers in the cross-section experience the full stress. The mean cross section stress is therefore less than the recorded stress.

(ii) In tensile loading, there is the possibility that the applied load is not exactly axial (i.e., the load may be eccentric). Thus we have some extra bending stress due to the eccentricity superimposed on the axial stress. Hence the actual stress may be higher than the recorded stress. (Refer to any standard book on strength of materials for the effects of an eccentrically applied axial load.)

To correct for the type of loading, a loading correction factor, $C_L$ is applied. For tensile loading, $C_L = 0.85$, hence $\sigma_e$ (axial load) = 0.85 x $\sigma_e$ (bending). And of course $C_L = 1$ if the load is bending.

## 10.4.2 Size of Specimen

A size effect is noticed in cases where there is a stress gradient. Figure 10.5 shows two specimens both loaded in bending to the same stress. At a distance, s, into the cross section, it is observed that the stress is higher in the larger specimen compared to the smaller one. To correct for effect of the size, a size correction factor, $C_s$, is applied whenever a component with a diameter more than 13 mm is loaded in bending or torsion. The mean value of $C_s$ is 0.85 for bending and torsion. For axial loading; $C_s = 1$ i.e., no size correction is required.

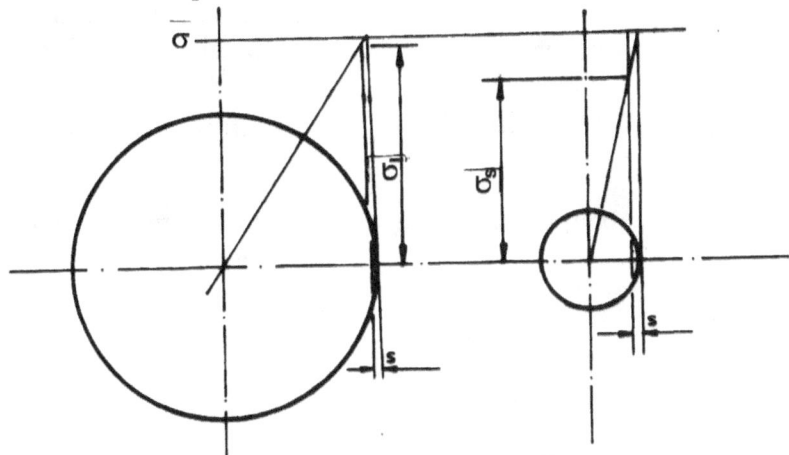

**Fig. 10.5** Size effect in fatigue

### 10.4.3 Surface Finish

As already stated, the endurance limit is determined in highly polished specimens. For specimens or components with a less perfect surface, the endurance limit is lower. The reason for surface finish effect is that fatigue life is composed of two parts: the number of cycles required to initiate a crack and the number of cycles to grow the crack to critical size. In a polished specimen, most of the life is taken by the initiation phase. In less perfect surfaces, the initiation life is considerably lower.

The effect of surface finish is not the same for all materials. It is highest in high strength steels, which have higher notch sensitivity. Little effect of surface finish is noticed in other materials like cast iron and non-ferrous alloys.

To correct for the surface finish, a surface finish factor, $K_s$, is applied. The variation of $K_s$ with ultimate tensile strength for steels is shown in figure 10.6.

### 10.4.4 Stress Concentration

The reader may be familiar with the concept of stress concentration from a study of strength of materials. In static loading, the geometric or Nueber's stress concentration factor $K_t$ is defined as:

$$K_t = \frac{maximum - local - stress}{nominal - stress} \qquad (10.4)$$

It is usually read from handbooks or charts. A similar effect is observed in fatigue and a fatigue strength reduction factor, $K_f$ can be defined as:

$$K_f = \frac{\sigma_e - without - notch}{\sigma_e - with - notch} \qquad (10.5)$$

$K_f$ depends on both $K_t$ and on the material's notch sensitivity index, q, defined as:

$$q = \frac{K_f - 1}{K_t - 1} \qquad (10.6)$$

Hence,

$$K_f = 1 + q(K_t - 1) \qquad (10.7)$$

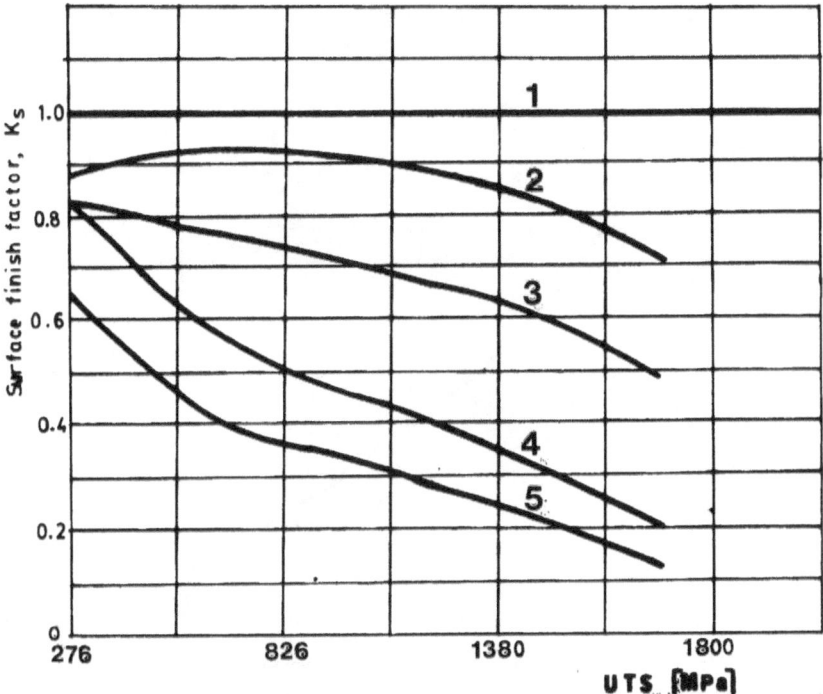

**Fig. 10.6** Schematic variation of surface finish factor with tensile strength in steels: (1) polished (2) ground (3) machined (4) rolled (5) forged.

q varies with the materials ultimate tensile strength as shown in figure 10.7 for steels.

## 10.4.5 Corrected Endurance Limit

In design against fatigue, a corrected endurance limit, $\sigma_e'$, can therefore be calculated from:

$$\sigma_{e'} = \frac{K_s \cdot C_L \cdot C_s}{K_f} \sigma_e \qquad (10.8)$$

where the factors $K_s$, $C_L$, $C_s$ and $K_f$ are as already defined above, while $\sigma_e$ is the intrinsic endurance limit determined as detailed in section 10.3. A modified S-N curve can then be constructed as shown in figure 10.8. Note that in the figure, one point corresponds to ($\sigma_{UTS}$, $10^3$). (Some authors give the point as ($0.9\sigma_{UTS}$, $10^3$). This is because in HCF, any value of N less than

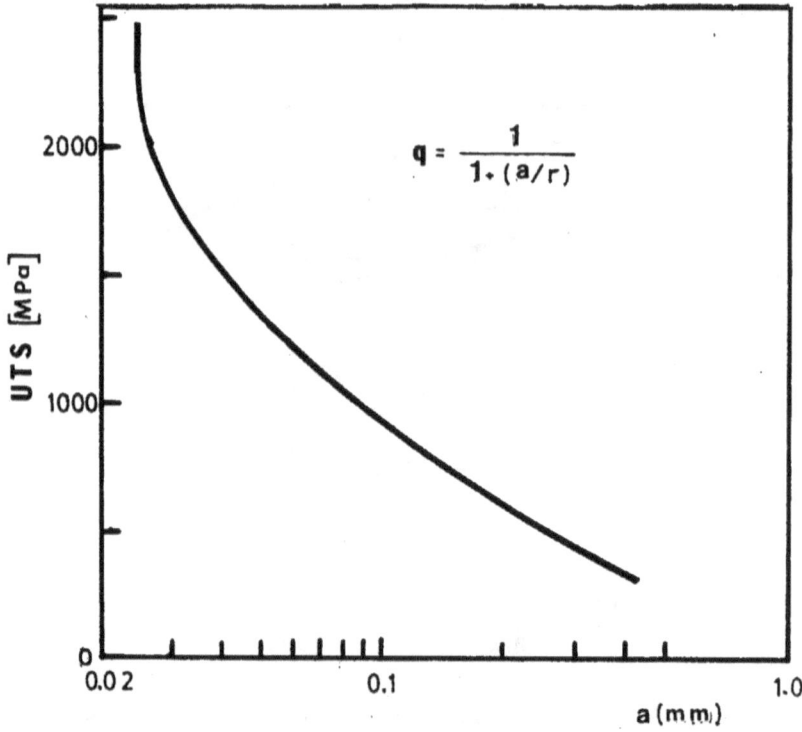

$$q = \frac{1}{1 + (a/r)}$$

**Fig. 10.7** Schematic variation of notch sensitivity with tensile strength.

$10^3$ is considered to be static loading. If the component does not fail on first application of the load, it will probably not fail in $10^3$ cycles). Since the factors considered in calculating the modified endurance limit do not affect the tensile strength, the modified and intrinsic curves coincide at this point. The other point is ($\sigma_e$, $10^7$) for the intrinsic graph, and ($\sigma_e'$, $10^7$) for the modified graph.

Example 10.1: Sketch, on the same diagram, the intrinsic and modified S-N curves for a steel with the following properties: Fatigue fracture strength at $10^3$ cycles = 600 MPa, endurance limit at $10^7$ cycles = 310, notch sensitivity index = 0.9; and diameter = 60 mm. The component in question is rolled and has a hole drilled through it causing a geometric stress concentration factor of 2.1. It is loaded in repeated bending. Estimate the life of the component under an alternating stress of 200 MPa, and under conditions similar to those given earlier.

154

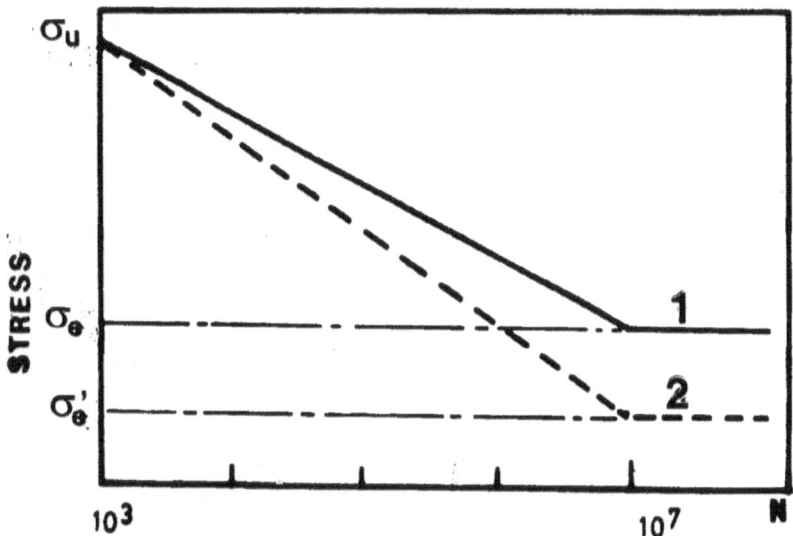

**Fig. 10.8** Intrinsic (1) and modified (2) S-N curves for steels

Answer:

It is first necessary to calculate the corrected endurance limit using equation 10.8:

$K_s = 0.58$ read from figure 10. 6 for a rolled surface

$C_L = 1$ since the load is a bending load

$C_s = 0.85$ since the bending load induces a stress gradient

$K_f$ is calculated from equation 10.7 substituting the given values of q and $K_t$

=> $K_f = 1 + 0.9(2.1 - 1) = 1.99$.

Then:

$$\sigma_{e'} = \frac{310x0.58x0.85}{1.99} = 77MPa$$

The S-N curves are as shown in figure E10.1. Drawing a horizontal line through $\sigma = 200$ MPa, and noting the value of N at the point it intercepts the modified S-N line, we determine the life for this stress as just over $10^6$ cycles.

## 10.5   EFFECTS OF MEAN STRESS
### 10.5.1 Goodman, Gerbers and Soderberg Formulae

In design against fatigue, it is the alternating stress that is compared with the endurance limit. But as can be expected, the allowable alternating stress will definitely be affected by the mean stress (see figure 10.2).

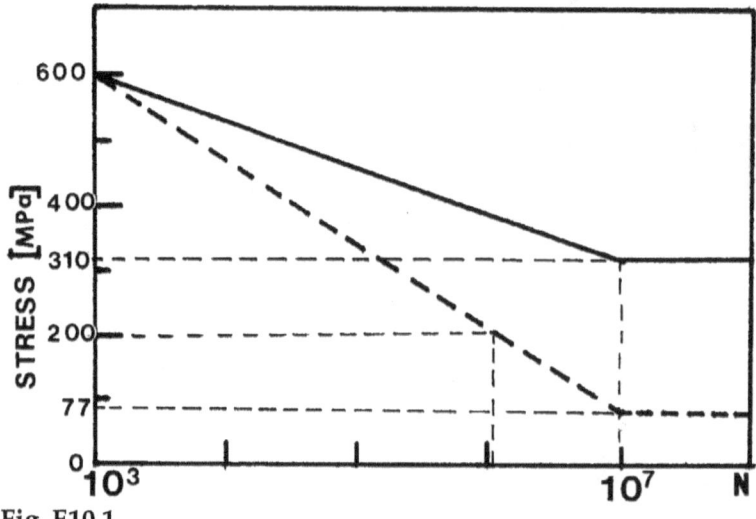

**Fig. E10.1**

Several equations have been proposed to take account of this mean stress effect. The most common are:

Gerbers formula:

$$\sigma_a = \sigma_e \left[ 1 - \left( \frac{\sigma_m}{\sigma_{UTS}} \right)^2 \right]$$
(10.9)

Goodman formula:

$$\sigma_a = \sigma_e \left[ 1 - \frac{\sigma_m}{\sigma_{UTS}} \right]$$
(10.10)

Soderberg formula:

$$\sigma_a = \sigma_e \left[ 1 - \frac{\sigma_m}{\sigma_Y} \right]$$
(10.11)

where $\sigma_a$ is the allowable alternating stress, $\sigma_m$ is the mean stress; $\sigma_Y$ is the yield stress while $\sigma_{UTS}$ is the ultimate tensile strength. These equations are also represented in the form of diagrams as shown in figure 10.9. Experimental results tend to agree most closely with the Goodman Equation.

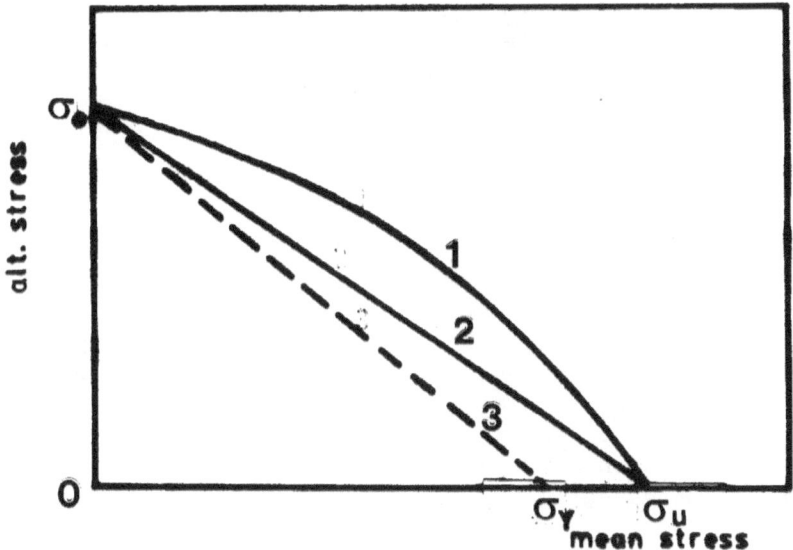

**Fig. 10.9** Effect of mean stress on endurance limit: (1) Gerber (2) Goodman (3) Soderberg

Example 10.2: Using the Goodman equation or diagram, determine the diameter of a steel shaft that can withstand a static tensile force of 36 kN and a cyclic load varying from zero to 72 kN when a factor of safety of 2 is required for both the mean and fluctuating components of the stress. The following have been determined: $\sigma_{UTS} = 670$ MPa; $\sigma_e = 335$ MPa; $K_s = 0.73$; $K_t = 2.02$; and $q = 0.86$.

Answer:
First, the corrected endurance limit is determined:
  $C_L = 0.85$ since the alternating load is tensile
  $C_s = 1$ since there is no stress gradient, hence no size effect
  $K_f = 1 + q(K_t - 1) = 1 + 0.86(2.02 - 1) = 1.87$

Then from equation 10.8:

$$\sigma_{e'} = \frac{0.73x0.85}{1.87} 335 = 110 MPa$$

If P represents the load:

$$P_a = \frac{P_{max} - P_{min}}{2} = \frac{108 - 36}{2} = 36kN$$

And:

$$P_m = \frac{P_{max} + P_{min}}{2} = \frac{108 + 36}{2} = 72kN$$

Hence:

$$\sigma_a = \frac{4 P_a}{\pi d^2}$$

A similar equation can be written for the mean stress. Applying the Goodman equation 10.10 and taking account of the factors of safety:

$$\sigma_a = \frac{\sigma_{e'}}{2} [1 - 2\frac{\sigma_m}{\sigma_{UTS}}]$$  (A)

Substituting for $\sigma_a$ and $\sigma_m$ from above into equation (A):

$$\frac{4x36,000}{\pi d^2} = \frac{110x10^6}{2} \left[1 - \frac{2x4x72,000}{\pi d^2 x670x10^6}\right]$$

Solving for d:  d = 33 mm

## 10.5.2 The Smith Diagram

The Smith diagram (or modified Goodman diagram) shown in figure 10.10 is sometimes used to assess the effect of mean stress especially in those instances in which the mean and alternating stresses are applied independently. The diagram is plotted as follows:

(i) Draw the $\sigma_m$ and $\sigma_a$ axes (use same scale for both axes) and mark off $\sigma_{UTS}$, $\sigma_Y$, $\sigma_{e'}$ and $-\sigma_{e'}$ as shown.

(ii) Draw line AP at 45° to the horizontal hence mark point Q at the mid-point of PR.

(iii) Draw AQ to intersect the horizontal $\sigma_Y$-C line at B.

(iv) Draw ER to intersect the vertical $\sigma_Y$-C line at D.

ABCDE then marks the boundary of the Smith diagram.

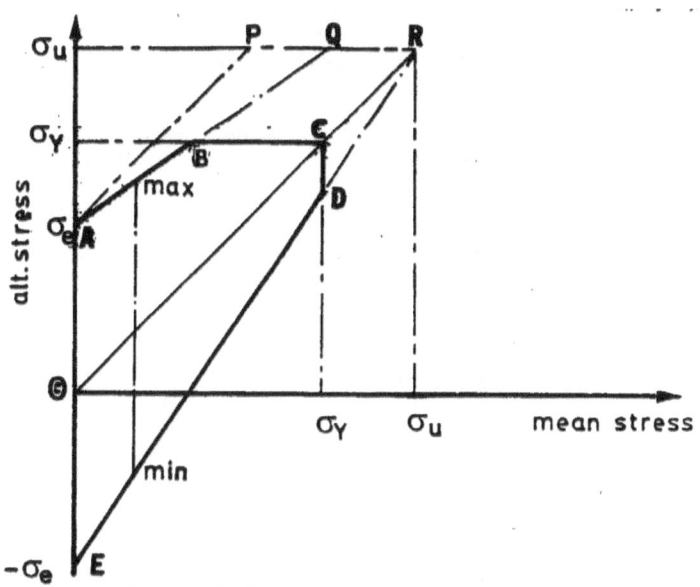

**Fig. 10.10** The Smith diagram

ABC gives the maximum stress allowable while DE gives the minimum stress allowable. To use the Smith diagram, draw a vertical line through the value of $\sigma_m$. The intersection of this vertical line with ABC and DE gives the values of $\sigma_{max}$ and $\sigma_{min}$ respectively. The allowable alternating stress is then determined from:

$$\sigma_{a.all} = \frac{\sigma_{max} - \sigma_{min}}{2} \qquad (10.12)$$

NOTE: For most applications, a negative mean stress is usually taken as being equal to zero mean stress if its magnitude is less than the yield stress in compression.

Example 10.3: A high strength steel has the following properties: tensile strength 1100 MPa, yield stress = 800 MPa, endurance limit at $10^7$ cycles =

600 MPa, $K_t$ = 1.5. The member has a ground surface and is loaded in pulsating tension with a mean stress of 280 MPa. Use the Smith diagram to determine the alternating stress allowable on the member if the smallest radius on the component is 3 mm.

Answer:

First, the endurance limit is calculated from equation 10. 8:

$K_s$ = 0.9 (from figure 10.6, line 2); $C_s$ = 1 (since the alternating load is an axial load); $C_L$ = 0.85 (since there is no stress gradient); and $K_f$ = 1 + q($K_t$ - 1).

To get q, use figure 10.7. For a UTS of 1100 MPa, the value of a is 0.07. Then:

$$q = \frac{1}{1+\dfrac{a}{r}}$$

And, with r = 3 mm, q = 0.98. From this, $K_f$ = 1.49. Then:

$$\sigma_{e'} = \frac{600 \times 0.9 \times 0.85}{1.49} = 308 MPa$$

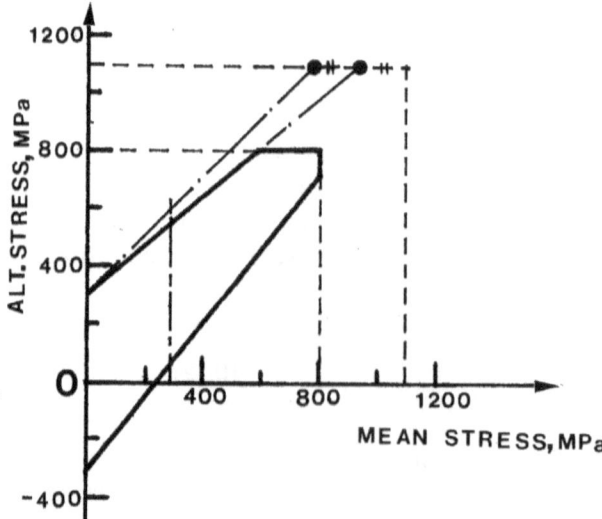

**Fig. E10.3**

The corresponding Smith diagram is shown in figure E10.3. From the diagram, values of maximum and minimum stress corresponding to a mean stress of 280 MPa are: Maximum stress = 540 MPa, minimum stress = 40 MPa. From equation 10.12, the allowable alternating stress is: $\sigma_{a.all}$ = 250 MPa.

## 10.6   CUMULATIVE FATIGUE

In most practical situations, the stress range is not constant and we need a method to assess the integrity taking into account the variation of stress amplitudes. Several formulae have been suggested but the most frequently used is the linear cumulative damage rule or Palmgren--Miner rule. The basis of the rule is that if the life of a component at a stress amplitude of $\sigma_1$ is $N_1$ and it endures $n_1$ cycles at this stress level, damage equivalent to $n_1/N_1$ is done. Similarly, damage equal to $n_2/N_2$ is done at $\sigma_2$. These damages add up or cumulate and when the sum of the damage = 1, failure occurs, (see figure 10.11) i.e., failure occurs when:

$$\sum_i \frac{n_i}{N_i} \geq 1 \qquad\qquad (10.13)$$

This rule is not always followed possibly because: (i) the sequence of the loads is not considered important by the rule; (ii) the phases of the varying loads are not taken into account.

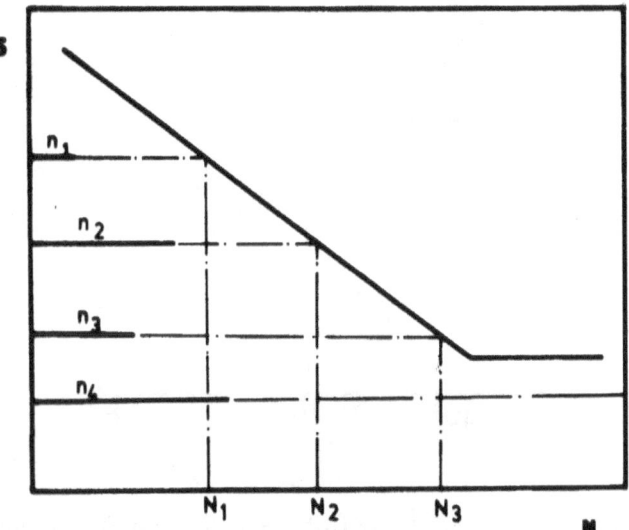

**Fig. 10.11** Illustration of Palmgren--Miner rule

Example 10.4: A 10 mm steel cable is used in a crane that lifts loads of 1600 kg, 1850 kg, and 2000 kg successively. On average, this load sequence is repeated 156 times a day, every day for five years before the cable is replaced. Use the Palmgren-Miner rule to establish if the cable is likely to fail before it is replaced. The steel has the following properties: endurance limit at $10^7$ cycles = 140 MPa, fatigue fracture strength at $10^3$ cycles = 450 MPa.

Answer:

Let the stress induced into the cable by the 1600 kg weight = $\sigma_1$. Then:

$$\sigma_1 = \frac{4x\,P_1\,g}{\pi\,d^2} = \frac{4x1600X9.81}{\pi(0.01\,)^2} = 200MPa$$

where g is the gravitational acceleration. Similarly, $\sigma_2$ = 230 MPa and $\sigma_3$ = 250 MPa.

$n_1 = n_2 = n_3$ = 156 times per day x 365 days per year x 5 years = 284 700 load cycles.

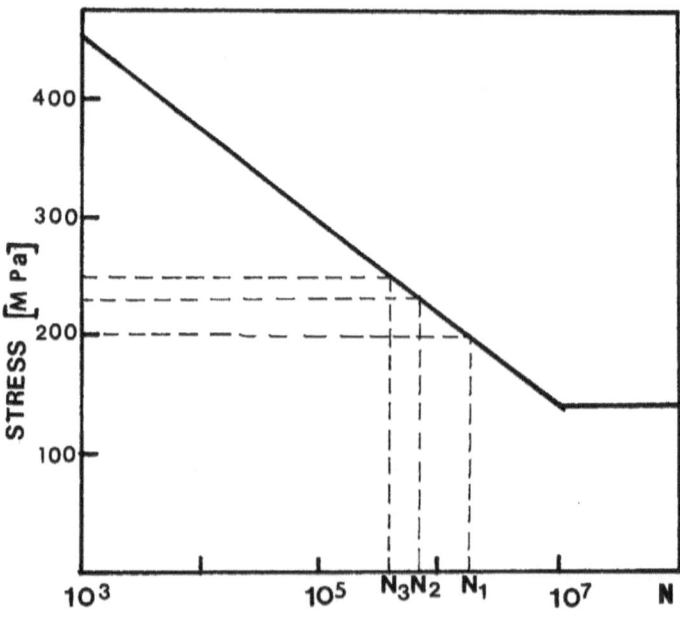

**Fig. E10. 4**

Next, the S-N curve is plotted (figure E10.4) and $N_1$, $N_2$ and $N_3$ obtained as the lives of the steel at the corresponding values of stress. From the diagram, $N_1 = 1.6 \times 10^6$, $N_2 = 6.7 \times 10^5$ and $N_3 = 3.8 \times 10^5$. Then:

$$\sum_i \frac{n_i}{N_i} = 2.84 \times 10^5 \left[\frac{1}{1.6 \times 10^6} + \frac{1}{6.7 \times 10^5} + \frac{1}{3.8 \times 10^5}\right] = 1.39$$

Since this total is greater than one, failure is likely to occur before replacement.

## 10.7    LOW CYCLE FATIGUE

As has been said already, low cycle fatigue refers to those loading situations where the stresses are high enough to produce considerable macroscopic plasticity. Comparatively fewer stress cycles are required to cause failure. Moreover, stress is no longer proportional to the strain and hence the strain range, has to be used for analysis. Low cycle fatigue analysis is particularly useful for the analysis of pressure vessels (nuclear and ordinary), thermal fatigue, landing gear of aircraft, etc. The following definitions, illustrated in figure 10.12 are relevant to the analysis of LCF:

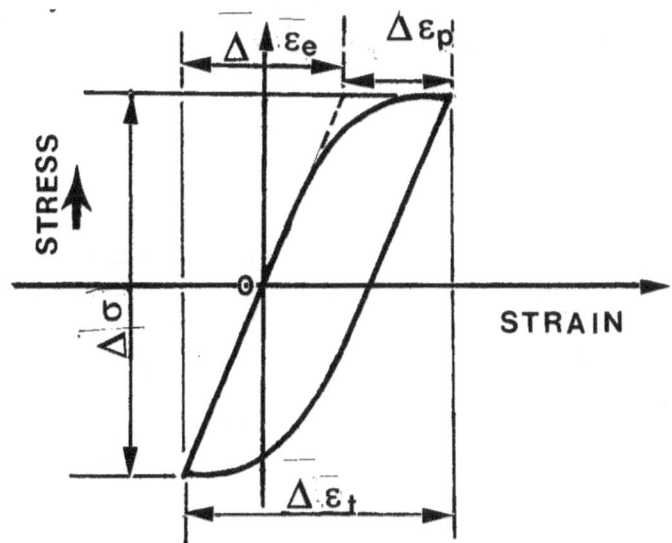

**Fig. 10.12** Definition of terms in low cycle fatigue

(i) $\Delta\varepsilon_e$ = elastic strain range = $\Delta\sigma/E$. The elastic strain amplitude is half this value.

(ii) $\Delta\varepsilon_p$ = plastic strain range. The plastic strain amplitude is half this value.

(iii) $\Delta\varepsilon_t$ = total strain range = $\Delta\varepsilon_e + \Delta\varepsilon_p$.

(iv) $\Delta\sigma$ = total stress range.

(v) N = number of stress reversals. Component life = 2N cycles.

For most materials, the relationship between the stress and strain in the plastic range is given by the Hollowman equation, i.e:

$$\sigma = K\,\varepsilon_p^n \qquad\qquad (10.14)$$

where K is the strength coefficient, and n is the strain hardening exponent. A similar equation may be written for cyclic loading as:

$$\Delta\sigma = K'\Delta\varepsilon_p^{n'} \qquad\qquad (10.15)$$

Here, K' is the cyclic strength coefficient, n' is the cyclic strain hardening exponent, and $\Delta\varepsilon_p$ is the cyclic plastic strain range.

It is customary in LCF to treat the elastic and plastic components separately and to relate each of these to the component life, 2N separately. Empirically, it has been determined that when the plastic strain amplitude, $\Delta\varepsilon_p/2$, is plotted against component life, 2N on log scales, a straight line is obtained. This suggests a relationship of the form:

$$\frac{\Delta\varepsilon_p}{2} = \varepsilon_f{}'(2N)^c \qquad\qquad (10.16)$$

where, $\varepsilon_f'$ is the fatigue ductility coefficient and is approximately equal to the material's true strain at fracture, $\varepsilon_f$. c is the fatigue ductility exponent which has been shown by experiment to vary between -0.7 and -0.5 for steels. Equation 10.16 is termed the Coffin-Manson equation after its proponents. The true strain at fracture can be shown (see the first reference at the end of this chapter) to be related to the materials percent reduction in area, $\psi$, by:

$$\varepsilon_f = \ln\frac{1}{1-\psi} \qquad\qquad (10.17)$$

A similar relationship exists between the elastic strain amplitude $\Delta\varepsilon_e/2$ and the component life, 2N i.e:

$$\frac{\Delta\varepsilon_e}{2} = \frac{\sigma_{f'}}{E}(2N)^b \qquad\qquad (10.18)$$

where $\sigma_f'$ is the fatigue strength coefficient, and is approximately equal to the material's true fracture stress. b is the fatigue strength exponent, which has been shown by experiment to lie between -0.05 to -0.12 for steels. The total strain range can then be expressed as:

$$\frac{\Delta\varepsilon_t}{2} = \frac{\sigma_{f'}}{E}(2N)^b + \varepsilon_f'(2N)^c \qquad\qquad (10.19)$$

The relationships above are shown graphically in figure 10.13. When cyclic material properties are not available, the following empirical approximation, due to Manson, may be used to estimate the total strain range corresponding to a given life:

$$\Delta\varepsilon_t = 3.5\frac{\sigma_{UTS}}{E}N^{-0.12} + \varepsilon_f^{0.6}N^{-0.6} \qquad\qquad (10.20)$$

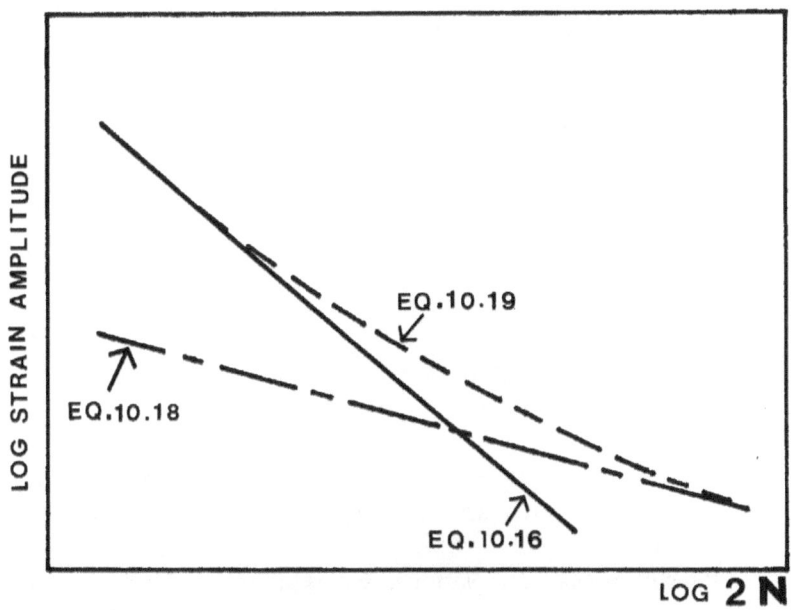

**Fig. 10.13** Relation between strain amplitude and life in LCF

Figure 10.13 shows, as expected, that the plastic component dominates at low values of N (high stress), while the elastic component dominates at high values of N.

Example 10.5: A turbine experiences a thermal cycle from room temperature (20 °C) to the service temperature (540 °C). The blades of the turbine are made from a material with the following properties: ultimate tensile strength = 550 MPa, yield stress = 410 MPa, percent reduction in area = 30 %, E = 165 GPa, coefficient of thermal expansion, $\alpha$ = 1.8 x 10⁻⁵/°C. How many thermal cycles may the blades withstand?

Answer:
$$\Delta\varepsilon_t = \alpha(T_s - T_r) = 1.8 \times 10^{-5} \times (540 - 20) = 9.36 \times 10^{-3}$$
At yielding: $\varepsilon_Y = \sigma_Y/E = 410/165\ 000 = 2.48 \times 10^{-3}$

From theses values, it can be appreciated that the strain amplitude is higher than the yield strain. At the same time, the strain amplitude must be less than the strain corresponding to the tensile strength. The corresponding stress amplitude must also lie between the yield stress and the UTS. By interpolation, an approximation of stress amplitude can be taken as the mean of the yield and ultimate stresses. Then:

$$\frac{\Delta\sigma}{2} = \frac{1}{2}(550 + 410) = 480 MPa$$

And

$$\frac{\Delta\varepsilon_e}{2} = \frac{\Delta\sigma}{2E} = 2.9x10^{-3}$$

$$\Rightarrow \Delta\varepsilon_e = 5.8 \times 10^{-3}$$
$$\Rightarrow \Delta\varepsilon_p = \Delta\varepsilon_t - \Delta\varepsilon_e = 3.56 \times 10^{-3}$$

Also, $\varepsilon_f'$ is approximately = $\varepsilon_f$, which may be calculated from the given value of percent reduction in area and equation 10.17.
$$\Rightarrow \varepsilon_f' = 0.36$$
From the coffin--Manson relationship and taking a value of c = -0.6

$$2N = \left[\frac{\varepsilon_p}{\varepsilon_f{}'}\right]^{1/-0.6} = 6968$$

From which:   N = 3 500 cycles

## 10.8   CORROSION FATIGUE

The term corrosion fatigue refers to the simultaneous action of cyclic load and a corrosive environment. This is found to result in a higher reduction in strength than that which would result if each of the factors acted independently. Under conditions of corrosion fatigue, no endurance limit is observed in any alloy and many more micro cracks develop at a much lower stress before one of them predominates. The greater damage observed in corrosion fatigue is thought to be due to the fact that the two processes assist each other: the cyclic loading assists corrosion by pumping the corroding medium in and out of the crack and hence renewing the medium and removing spent corrosion products. It (the cyclic load) also helps to break any protective layer that may have formed at the crack tip and hence exposes fresh metal to further attack. A corrosive environment assists the fatigue process by embrittling the crack tip hence making it easier for crack extension to occur. The actual mechanism of corrosion fatigue are however, complex and outside the scope of this book.

## 10.9   FRETTING

This refers to the damage caused due to slight relative movement between two (nominally) well mating parts (e.g., a hub force fitted to a shaft). It is a combination of mechanical and chemical effects in which metal is removed by grinding or welding and tearing. The particles removed act as abrasives hence causing more damage. Fretting damage is observed to be much more severe in presence of oxygen and nitrogen hence exclusion of air from the mating surfaces reduces fretting. The sites of fretting damage act as fatigue crack initiation sites thus reducing greatly the fatigue life.

Fretting can be prevented or reduced by doing the following:

1. Increasing the contact pressure between contact surfaces or otherwise preventing relative movement between them.

2. Reducing the friction between parts in contact e.g., by placing a material with less coefficient of friction between mating parts. This method is used to reduce fretting in leaf spring assemblies where hard plastics are

placed between the leaves.

3. Increasing the wear resistance of the surface in contact e.g., by case hardening or nitriding.

4. Excluding oxygen and nitrogen (air) from the contact surface e.g., by applying lubrication.

## 10.10  CONCLUDING REMARKS

Like all analysis involving growth of cracks, design against fatigue is nowadays carried out using the science of fracture mechanics. With this new method, the parameter stress is replaced by the stress intensity factor, K, which takes account of both the stress and the crack length. It is then possible to separate the initiation and propagation phases of fatigue and to calculate the rate of fatigue crack growth. In the region where the alternating stresses (and hence the stress intensity range, $\Delta K$) are moderate, the rate of crack growth per cycle, da/dN, can be approximated by:

$$\frac{da}{dN} = C(\Delta K)^m \qquad (10.21)$$

where C and m are constants. This equation is termed the Paris equation. Further consideration of fracture mechanics is however, outside the scope of the present book.

## PROBLEMS

10.1 Explain briefly why (i) the size of a specimen (ii) the surface finish, affect the fatigue strength of steels.

10.2 Sketch, on the same diagram, the intrinsic and modified fatigue curves of a steel shaft with the following properties: $\sigma_{UTS} = 800$ MPa; $\sigma_Y = 550$ MPa; $\sigma_e = 380$ MPa. The solid shaft, 30 mm diameter is un-notched and has a machined surface. It is loaded in pulsating tension.

In another application, the same size of shaft is loaded by a static tensile force of 250 kN. Using the Smith diagram, determine the maximum completely reversed bending force, P, allowable at mid-span for a span of 2 m. (Ans: $\sigma_e' = 239$ MPa; 835 N.)

10.3 A crane in a busy factory has a steel cable with the following properties: fatigue fracture stress at $10^3$ cycles = 420 MPa, endurance limit at $5 \times 10^6$ cycles = 150 MPa. The crane is used to lift three different loads which induce the following respective stress levels in the cable: $\sigma_1$ = 203 MPa, $\sigma_2$ = 232 MPa, and $\sigma_3$ = 255 MPa. On average, the crane is assumed to lift each of the three loads 56 times daily for 5 years continuously before the cable is replaced. Check whether, based on the Palmgren-Miner rule, failure is likely to occur before the cable is replaced. (Ans: Yes).

10.4 A steel member with the following properties: $\sigma_{UTS}$ = 400 MPa, $\sigma_Y$ = 220 MPa, $\sigma_e$ = 170 MPa, is loaded by a mean stress of 120 MPa. Determine the allowable alternating stress based on the following relationships: (i) Soderberg (ii) Goodman (iii) Gerber. (Ans: 77 MPa, 119 MPa, 155 MPa.)

10.5 An un-notched steel shaft is loaded as shown schematically in figure Q10.5. If the stress induced is moderate, sketch the fatigue fractured surface expected from each loading.

10.6 Using the Goodman equation, determine the diameter of a steel shaft that can withstand a static tensile force of 36 kN and a dynamic tensile force varying from 0 to 72 kN when a safety factor of 2 is desired for both the fluctuating and mean components of the stress. From the material used

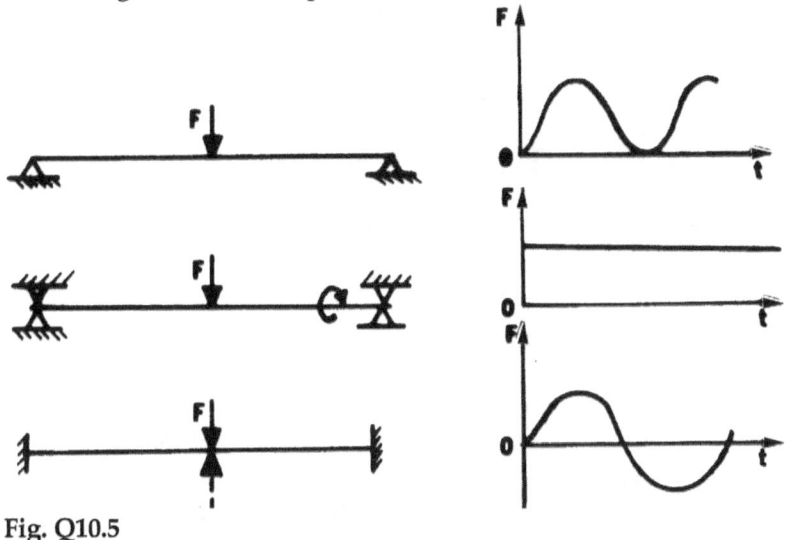

**Fig. Q10.5**

for the shaft and the design, the following have been determined: $\sigma_{UTS}$ = 670 MPa, $K_t$ = 2.02, fillet radius r = 4.8 mm, q = 0.86, $\sigma_e$ = 335 MPa, $K_s$ = 0.73. (Ans: 33 mm.)

10.7 The flat rectangular beam shown in figure Q.10.7 (a) is to be used as a spring with the force applied at the end as shown. The force cycle shown in figure Q.10.7 (b) is to be repeated 280,000 times. Check whether, based on the Palmgren-Miner rule, failure is likely to occur due to cumulative fatigue. The properties of the material are: failure strength at $10^3$ cycles = 600 MPa, endurance limit at $10^7$ cycles = 280 MPa. (Ans: Yes.)

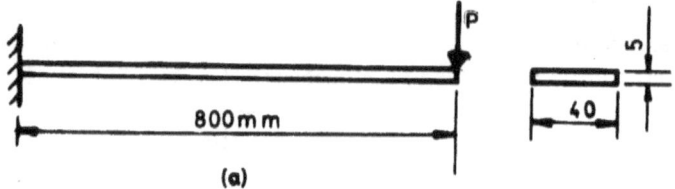

(a)

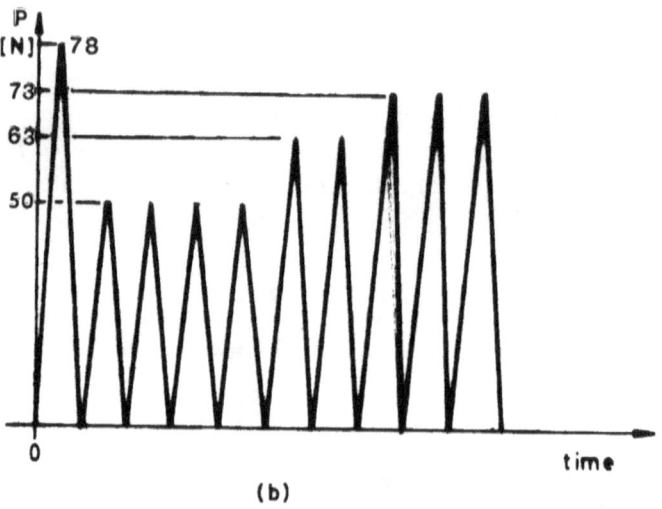

(b)

**Fig. Q10.7**

10.8 Explain what is meant by low cycle fatigue (LCF) and why the stress range cannot be reliably used for its analysis. Give one example where LCF analysis may be employed.

The fatigue strength of an aluminium alloy may be represented by the following equation:

$$[\sigma_r]^b [N_f] = C$$

where $\sigma_r$ is the stress range, $N_f$ is the number of cycles to failure and b and C are empirical constants. It is determined that the cyclic yield stress of the alloy is 400 MPa while the ultimate tensile strength is 550 MPa. When the stress range = 340 MPa, $N_f = 10^8$ cycles whereas a stress range of 520 MPa corresponds to $N_f = 10^5$ cycles.

A component is to be machined from the alloy, so as to have a life of $10^7$ cycles when a mean stress of 200 MPa is applied. Using the Smith diagram, determine the allowable alternating stress. (Ans: 125 MPa.)

10.9 A 50 mm square bar is loaded by a static tensile force, $P_s = 200$ kN. The bar is machined from steel with the following properties: $\sigma_{UTS} = 650$ MPa, $\sigma_Y = 420$ MPa, $\sigma_e = 300$ MPa, $K_s = 0.75$.

The bar is supported at points 3 m apart (assume simple supports for simplicity) and a completely reversed bending force, $P_b$ applied at a point 1 m from one of the supports. Use the Smith diagram to calculate the value of $P_b$ to avoid fatigue failure. (Ans: 5.3 kN.)

10.10 Under what loading conditions would you expect the size of a specimen to have an effect on the endurance limit in fatigue and why?

10.11 A 10 mm steel cable is used in a crane at the Kilindini Habour. The crane lifts loads of 1100 kg, 1600 kg, 1800 kg, and 2000 kg successively. This load sequence is repeated 156 times every day for five years before the cable is replaced. Find out whether, based on the linear cumulative damage rule, the cable is likely to fail before it is replaced. The steel has the following properties: fatigue fracture strength at $10^3$ cycles = 450 MPa, endurance limit at 2 x $10^7$ cycles = 140 MPa. (Note: no correction on the endurance limit is required.) (Ans: No.)

10.12 What is cumulative fatigue? Describe one method that is most often used for the analysis of cumulative fatigue and state two reasons why the predictions of the method are not always followed in practice.

10.13 An un-notched solid steel shaft, 30 mm in diameter with a machined surface ($K_s = 0.74$) has the following properties: $\sigma_{UTS} = 800$ MPa, $\sigma_Y = 550$ MPa, $\sigma_e$ (at $10^7$ cycles) = 380 MPa. If the shaft is loaded in pulsating tension, estimate its life under an alternating stress of 490 MPa. (Ans: 1.7 x $10^5$ cycles.)

10.14 Using the Soderberg equation (or diagram), determine the diameter of a steel shaft that can withstand a static tensile force of 30 kN and a dynamic tensile force varying from 20 kN to 100 kN. The properties of the material are: $\sigma_{UTS}$ = 670 MPa, $\sigma_Y$ = 520 MPa, $\sigma_e'$ = 230 MPa, factor of safety on alternating stress = 2.0, factor of safety on mean stress = 1.4. (Ans: 27.4 mm.)

10.15 A 60 mm diameter steel shaft is loaded by a static tensile force of 300 kN. The bar is machined from steel with the following properties: $\sigma_{UTS}$ = 720 MPa, $\sigma_Y$ = 550 MPa, $\sigma_e'$ = 360 MPa. The bar is held as a cantilever and a completely reversed force of 3.8 kN applied at the end. Use the Smith diagram to determine the allowable length of the bar if fatigue failure is to be avoided. (Ans: 1.76 m.)

10.16. A solid circular shaft with a diameter of 70 mm is supported between bearings 1.6 m apart. A gear is keyed to the shaft 0.6 m from one bearing. The shaft runs continuously at 80 rpm while the load applied at the gear varies as follows: 6 days: 34 kN; 2 days: 42 kN; 1 day: 48 kN. Use the Palmgren-Miner rule to estimate how long the shaft can then operate when the load on the gear is 45 kN. You may treat the bearings as simple supports and the load as a point load. The material of the shaft has the following properties: fatigue fracture strength of 650 MPa at 1000 cycles and an endurance limit (corrected) of 300 MPa at $5 \times 10^7$ cycles. (Ans: Infinite.)

10.17. Sketch, on the same graph, the intrinsic and modified fatigue curves for a steel with the following properties: diameter 60 mm; fatigue failure strength at $10^3$ cycles; 600 MPa; endurance limit at $10^7$ cycles; 310 MPa; q = 0.6; $\sigma_{UTS}$ = 650 MPa. The member has a hole drilled in it giving $K_t$= 2.1. It is loaded in repeated bending. (Ans: $\sigma_e'$ = 77 MPa.)

10.18 (a) Explain what the following terms mean: (i) corrosion fatigue (ii) fretting.

(b) A high strength steel member has the following properties: $\sigma_{UTS}$ = 1100 MPa, $\sigma_Y$ = 800 MPa, $\sigma_e$ = 550 MPa, $K_t$ = 1.8, minimum radius = 3 mm. The member is machined and loaded in pulsating tension. A mean stress of 280 MPa is applied. Using the Smith diagram find the alternating stress allowable on the member. (Ans: 140 MPa.)

10.19 In a cyclic stress-strain curve, the stress amplitude is 75 MPa, the total strain amplitude is $6.45 \times 10^{-4}$, the true strain at fracture = 0.3, and the modulus of elasticity = 220 GPa. Determine the elastic and plastic strain ranges and the number of stress cycles likely to cause failure. (Ans: $6.8 \times 10^{-4}$, $6.1 \times 10^{-4}$, 49 000.)

10.20 Use Manson's approximation to determine the total cyclic strain range permissible in a component required to withstand $4.9 \times 10^4$ stress cycles if it is made from a material with the following properties: tensile strength = 350 MPa, Young's modulus = 207 GPa, yield stress = 200 MPa, and the true strain at fracture = 0.3. (Ans: $2.36 \times 10^{-3}$.)

10.21 For the turbine in example 10.5 above, calculate the allowable operating temperature if it has to operate for $4.9 \times 10^3$ cycles. (Ans: 498 °C.)

10.22 A thin walled pressure vessel handling some hot liquid is required to withstand 3500 cycles of operation. When the vessel is emptied, both the temperature and the pressure fall to room temperature values (pressure = 0, temperature = 20 °C). The temperature of the liquid is 540 °C, while the maximum pressure is 20 MPa. Determine the required wall thickness if the mean diameter is 1 m. The properties of the material are: yield stress = 410 MPa, tensile strength = 550 MPa, E = 170 GPa, percent reduction in area = 30 %, and coefficient of thermal expansion = $1.8 \times 10^{-5}$/K. (Ans: 20 mm.)

10.23 Estimate the number of stress cycles likely to cause failure in a superalloy component made from a material which has the following properties: tensile strength = 600 MPa, yield stress = 480 MPa, E= 170 GPa, percent reduction in area = 28 %, and coefficient of thermal expansion = $1.8 \times 10^{-5}$/°C. Use Manson's approximation to estimate the maximum temperature of operation if the component must withstand $10^4$ stress cycles. (Ans: 2960, 361 °C.)

## Further Reading

Frost, N. E., Marsh, K. J. and Pook, L. P. Metal Fatigue, Clanderon Press, 1974.

Duggan, T.V. and Byrne, J. Fatigue as a Design Criterion, McMillan Press,

1979.

Shigley, J. E. <u>Mechanical Engineering Design</u>, McGraw-Hill, New York, 1986.

# CHAPTER ELEVEN

# CREEP AND CREEP FAILURES

## 11.1    INTRODUCTION

The term creep refers to a time dependent increase in strain at a constant stress. The phenomenon is important in visco-elastic materials (polymers) and becomes important in metals if they are operating at high temperatures. Thus creep failure has to be considered in components like turbines, rocket engines, missiles, etc., that operate at high temperatures.

Creep may lead to fracture or may constitute a failure before fracture takes place. For example a turbine blade, which through creep, becomes too long as to touch the casing will have failed even though no fracture has taken place. Creep takes place at stresses much lower than the yield stress of the material.

Whether a material creeps or not is controlled by the relationship between its melting point and the operating temperature. A ratio, termed the homologous temperature may be defined as the operating temperature divided by the melting temperature (both on the absolute scale). Creep becomes important if the homologous temperature is higher than 0.3.

## 11.2    THE CREEP CURVE

The usual method adopted to determine the creep properties of a material is to apply a constant load to the specimen maintained at the temperature of interest. The strain is then recorded as a function of the time. The general shape of the resulting curve is as shown in figure 11.1. The figure is seen to have four distinct stages:

(i) $\varepsilon_o$: Represents the instantaneous strain recorded immediately the load is applied. Since most tests are carried out with stresses well below yield stress, $\varepsilon_o = \sigma_o / E$, where E is the modulus of elasticity.

(ii) Stage I: This stage is termed the stage of primary creep. It is observed that the rate of creep (gradient of the curve in figure 11.1) decreases with the amount of creep. It is a transient stage and the increased resistance is due to effects of strain hardening.

(iii) Stage II: The secondary creep stage. A constant creep rate is observed due to a balance between the effects of strain hardening and recovery. The creep rate during this state is the lowest along the whole curve and is what is referred to as the "creep rate". This stage is usually the most important for engineering purposes.

(iv) Stage III: The tertiary stage. This stage only manifests itself for tests carried out at constant load. The increased creep rate in this stage results from necking of the specimen, which causes an increase in the true stress. If the test were carried out at constant stress, a continuation of the secondary stage would be recorded as shown by the dotted line in figure 11.1.

The creep curves are affected by the stress and temperature as shown in figure 11.2. From the figure, it may be noted that increasing the stress or test temperature has the following effects:

(i) The instantaneous strain, $\varepsilon_0$ is increased.

(ii) The creep rate at the secondary stage increases.

(iii) The overall time to fracture decreases.

(iv) The strain at fracture may be higher.

(v) At high stress or temperature, the secondary stage cannot be distinguished any more, while at sufficiently low stress or temperature, the secondary stage is horizontal i.e., no further increase in strain takes place.

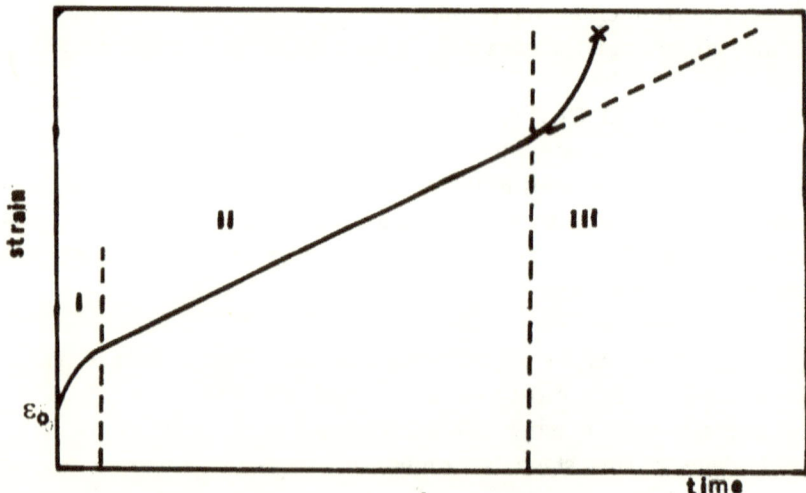

**Fig. 11.1** The creep curve

## 11.3   REASONS FOR CREEP

The following reasons explain the occurrence of creep at high homologous temperatures:

(a) High temperature causes an increase in the mobility of atoms and vacancies, which results in an increase in the diffusion rate and hence creep.

(b) The number of vacancies increases with increase in temperature. Since vacancies aid diffusion, creep becomes easier.

(c) Slip becomes easier at high temperature. Moreover, other slip systems, which are dormant at low temperature, may become active when the temperature is raised. Easy slip means easier deformation leading to creep.

(d) High temperature may cause changes in microstructure e.g., in steels, cementite may break down and eventually dissolve. In precipitation strengthened alloys, precipitate dissolution or over-aging may occur. These result in a weakening of the material, easier deformation, and hence creep.

(e) High temperature may also result in oxidation and intergranular corrosion and cause deformation at grain boundaries.

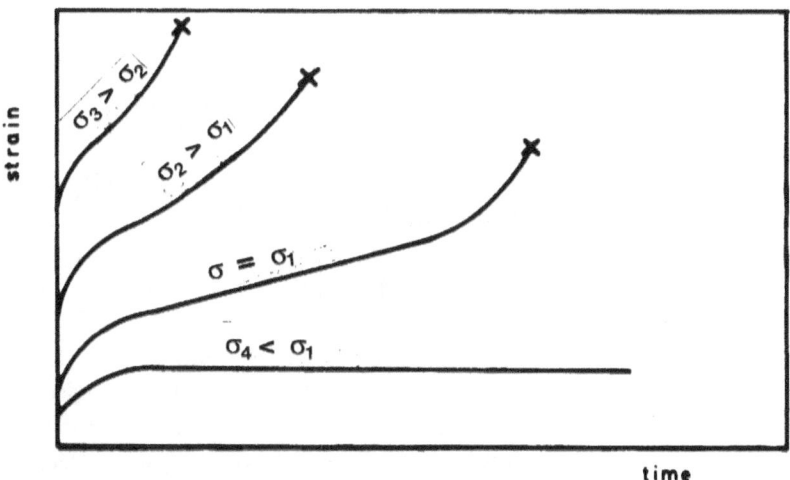

**Fig. 11.2** The effect of stress on the creep curve

## 11.4 CREEP MECHANISMS

The question to be answered is how, on an atomic scale, does the deformation responsible for creep take place? Creep can proceed by several mechanisms. These mechanisms in general operate independently of each other and the overall creep rate is determined by the rate of the fastest mechanism. The dominant mechanism depends on the applied stress and the temperature. In general, the mechanisms can be divided into two broad categories: mechanisms dependent on the movement of dislocations, and those dependent on the diffusion of atoms and vacancies.

## 11.4.1 Dislocation motion

This mechanism operates when the applied stress is relatively high, i.e., of the order of $10^{-4}$ times the shear modulus, G. The elevated temperature releases dislocations that might have been blocked by obstacles by causing them to climb to another plane with lower energy barriers. The rate of creep is represented by:

$$\frac{d\varepsilon}{dt} = \frac{ADGb}{kT}(\frac{\sigma}{G})^n \qquad (11.1)$$

where A is a constant, D is the diffusivity, G is the shear modulus, **b** is Burger's vector, k is Boltzmann's constant, T is the temperature on absolute scale while $\sigma$ is the stress applied. If the temperature at which creep takes place is constant, this may be written as:

$$\frac{d\varepsilon}{dt} = A'\sigma^n \qquad (11.2)$$

where A' is a constant. Hence, a plot of strain rate against applied stress on logarithmic scales will be a straight line. Due to the dependence on stress raised to a power, this creep mechanism is sometimes termed power law creep. Experimentally, it has been determined that the slope of the curve changes as shown in figure 11.3.

In the low stress region, n is approximately equal to one. This sub-region is termed the Harper-Dorn creep. In the next region, n is approximately equal to five. This is the power law creep proper. At still higher stress, the slope increases even more. This sub region is referred to as power law breakdown. The main determining factor governing the processes is the activation energy required to overcome obstacles to dislocation movement.

## 11.4.2 Diffusion Mechanisms

Diffusion mechanisms become dominant at high homologous temperatures in combination with low stresses. The principal reason for this is the higher mobility of atoms and vacancies at the high temperatures.

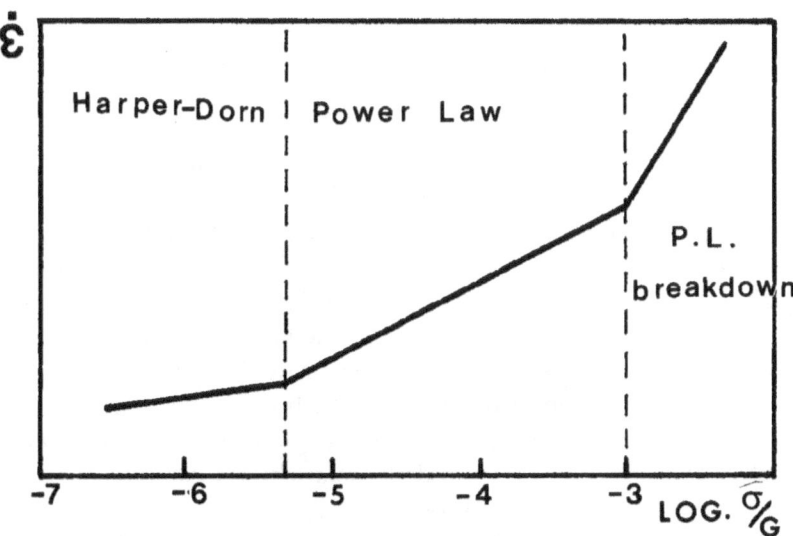

**Fig. 11.3** Various sub stages of dislocation creep

They involve the diffusion of vacancies and atoms. The diffusion may proceed either through the grain, or along the grain boundary. The mechanism involving diffusion of vacancies and atoms through the grain (also called lattice diffusion) is termed Nabarro-Herring creep. Vacancies diffuse from those grain boundaries under high stress to those under low stress. Atoms diffuse in the opposite direction. The net result is an elongation of the grain, which accounts for the creep deformation as shown in figure 11.4.

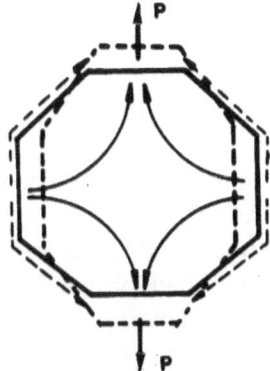

**Fig. 11.4** Lattice diffusion in creep

The creep rate under this mechanism is given by:

$$\frac{d\varepsilon}{dt} = \frac{A\sigma b^3 D_o}{kTd^2} \exp(\frac{-Q}{kT})  \qquad (11.3)$$

where, $D_o$ is the coefficient of lattice diffusion, d is the grain size, Q is the activation energy for lattice diffusion while the other terms have the same meaning as already defined. Equation 11.3 indicates that the rate of diffusion is inversely proportional to the square of the grain diameter. This explains why finer grained materials have lower creep resistance. It also explains why products made for high temperature applications, e.g., turbine blades are made as single crystals or unidirectionally cast crystals. This behavior is the opposite of the Hall-Petch Effect where the yield stress of a material is known to increase when the grain size is reduced, and also the effect of grain size on high cycle fatigue where a decrease of grain size results in an increase in the endurance limit.

At temperatures lower than those required for Nabarro-Herring creep, diffusion takes place along dislocation cores and grain boundary leading to the mechanism termed Coble creep. The creep rate is given by:

$$\frac{d\varepsilon}{dt} = \frac{A\sigma b^4 D_{gb}}{kTd^3}  \qquad (11.4)$$

where $D_{gb}$ is the coefficient of diffusion through the grain boundary. Note that Coble creep is inversely proportional to the cube of the grain diameter, and that $D_{gb}$ may be expressed by a relationship of the type:

$$D_{gb} = D_o \exp(-Q/kT)  \qquad (11.5)$$

This shows that equations 11.3 and 11.4 are of the same form.

## 11.4.3   Deformation Mechanism Maps

The information on deformation mechanisms may be summarized in deformation mechanism maps first proposed by Ashby. These are maps on stress-temperature space showing conditions under which a specific mechanism is the rate controlling mechanism. A simplified form of Ashby's deformation mechanism map is shown in figure 11.5. The axes of the maps are normalized stress and normalized temperature (actually, the homologous temperature).

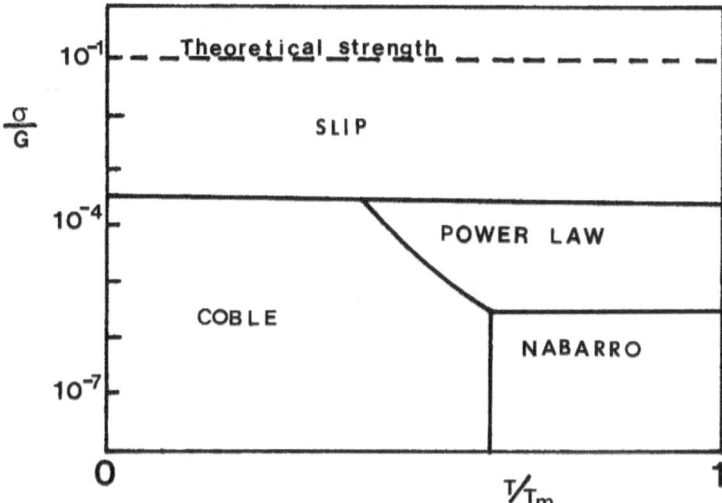

**Fig. 11.5** Schematic representation of a simplified deformation mechanism map

### 11.5  MECHANISMS OF CREEP RUPTURE

Eventually, creep leads to rupture. Three main mechanisms are responsible for creep rupture. These are transgranular creep, intergranular wedge cracking and formation of round type grain boundary cavities. Transgranular creep (figure 11.6 (a)) involves the nucleation and growth of cavities within the grain. The nucleation sites are inclusions within the grain which are detached from the matrix due to differences in Poisson's ratio between the inclusion and the matrix. Increased stress leads to local necking and hence coalescence of the cavities and final fracture. Intergranular wedge cracking or grain boundary triple point cracking is illustrated in figure 11.6 (b). It may be initiated by diffusion cavitation where diffusion of atoms from some part of the grain boundary leaves a cavity. The sliding of grain boundaries under the applied stress leads to a wedge type crack, which then spreads until fracture. Round type cavitation damage proceeds in a similar manner but with the grain boundary concerned perpendicular to the principal stress. The cavities coalesce by tensile fracture of the necks as illustrated in figure 11.6 (c). As in the case of creep deformation, creep rupture maps similar to figure 11.5 may be constructed for any specific alloy.

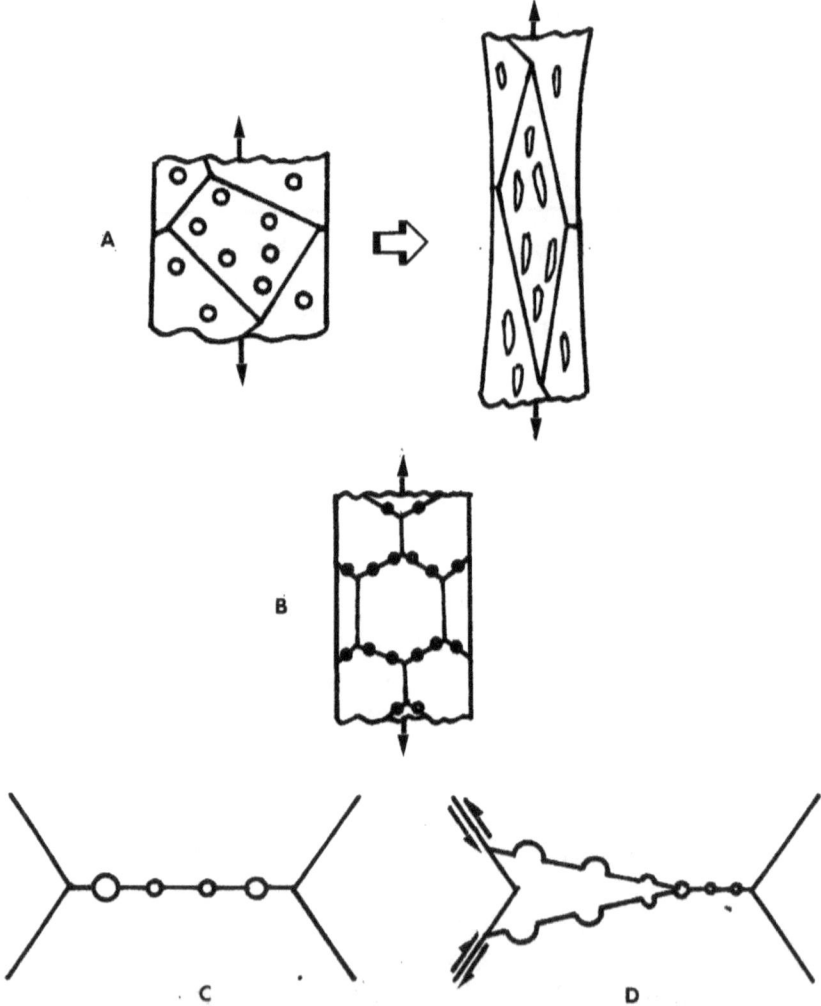

**Fig. 11.6** Mechanisms of creep rupture (A) transgranular creep (B) intergranular cracking (C) formation of r-type cavities (D) formation of w-type cavities.

## 11.6 DESIGN AGAINST CREEP

Eventually, the engineer is interested in producing a design that will not fail (due to creep). For this, a limiting stress is required. Failure in creep may either be fracture or excessive extension resulting from creep deformation. Material properties used in design against both types of failure are collected in two types of tests:

(i) Creep tests are performed to determine the creep strength of a material. This is the stress needed to produce a given strain rate (or a given strain in a given time). The creep strength is designated $\sigma_{x/h}$ where x is the limiting strain and h is the time to produce this strain. For example, if for a material $\sigma_{1/100,000} = 50$ MPa, a stress of 50 MPa will produce a strain of 1 % in 100,000 hrs. Creep strength tests are performed by accurate determination of the creep rate as a function of the stress (at a given temperature). Low stresses are used for such tests and hence long test times are involved. The strain (as a function of the time) is recorded very accurately and the tests are stopped at relatively low strains (of the order of 0.2 %).

(ii) Stress rupture tests: These are performed to determine a materials' rupture strength (i.e., results are used for design against rupture). The parameters recorded in such a test are the stress and the time to rupture (at a given temperature). Higher stresses and shorter test times are recorded for rupture tests compared to creep strength tests.

In both creep strength and creep rupture tests, the tests are repeated at different temperatures and the data plotted in diagrams similar to those shown in figure 11.7. It is clear that a lot more data must be collected due to the fact that three variables (time, temperature, stress) are involved. In design, the applied stress is limited to either $\sigma_{x/h}$ or the rupture strength depending on which type of failure is being considered. It is useful to ensure that the yield stress of the material (which reduces with increase in temperature) at the operating temperature is not exceeded.

Example 11.1: A tube, 80 mm external diameter, operates at 500 ℃ while under a pressure of 13 MPa. The strain in the material is limited to 1 % after 10 000 hrs. If the creep strength of the material at 500 ℃, $\sigma_{1/10\,000}$ is 140 MPa, determine the required wall thickness of the tube.

Answer:
For a thin walled tube, thickness, t, under internal pressure, p the maximum principal stress is given by:

$$\sigma = \frac{pd}{2t}$$

Making t the subject and substituting the given values:

$$t = \frac{pd}{2\sigma} = \frac{13x\,10^6\,x0.08}{2x140x\,10^6}\,m = 3.7mm$$

## 11.7  STRESS RELAXATION

If the extension of a material is constrained in conditions under which creep would have taken place, a stress relaxation (i.e., decrease of stress with time) takes place. Examples of components where stress relaxation may take place are bolts holding rigid plates, press fitted assemblies, etc. If the creep rate equation is known, it is easy to calculate the stress as a function of time as illustrated below. The total strain in the material, $\varepsilon_t$ is made up of an elastic component, $\varepsilon_e$, and a plastic component, $\varepsilon_p$, i.e:

$$\varepsilon_t = \varepsilon_e + \varepsilon_p = {}^{\sigma}/_E + \varepsilon_p \qquad (11.6)$$

If it is assumed that the material obeys a power law relationship between the strain rate and the stress, then:

$$\frac{d\varepsilon_p}{dt} = B\sigma^n \qquad (11.7)$$

If the strain is constant, $d\varepsilon_t/dt = 0$. Then:

$$\frac{d\varepsilon_t}{dt} = \frac{1}{E}\frac{d\sigma}{dt} + \frac{d\varepsilon_p}{dt} = \frac{1}{E}\frac{d\sigma}{dt} + B\sigma^n = 0 \quad (11.8)$$

Hence:

$$\frac{1}{E}\frac{d\sigma}{dt} = -B\sigma^n \qquad (11.9)$$

And:

$$\int\frac{d\sigma}{\sigma^n} = -BEf\,dt \qquad (11.10)$$

$$\frac{-1}{(n-1)\sigma^{n-1}} = -BEt + C \qquad (11.11)$$

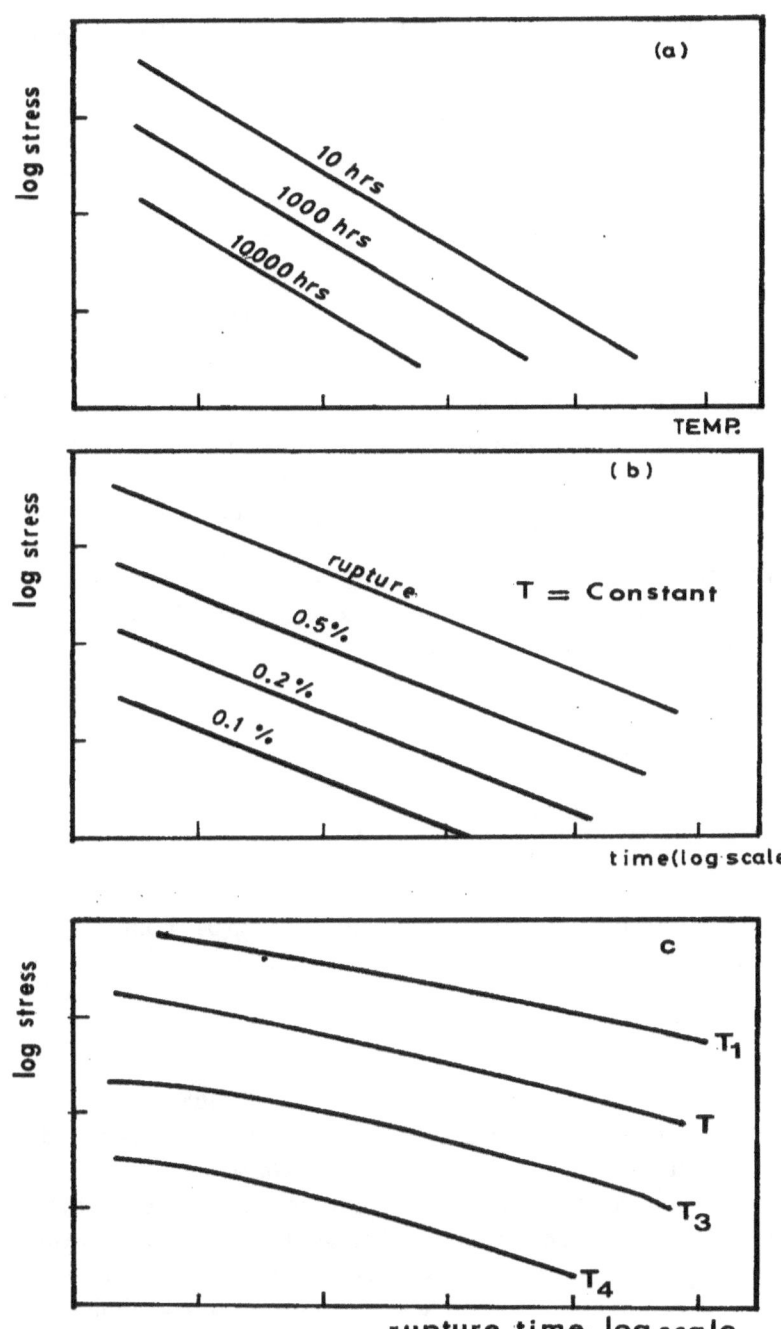

**Fig. 11.7** Various methods of representing creep data

At t = 0, stress = $\sigma_i$ = the initial stress. Substituting these boundary conditions into equation (11.11):

$$C = \frac{-1}{(n-1)\sigma_i^{n-1}} \qquad (11.12)$$

Substituting for C in equation (11.11):

$$\frac{1}{\sigma^{n-1}} = \frac{1}{\sigma_i^{n-1}} + (n-1)BEt \qquad (11.13)$$

Example 11.2: A 20 mm diameter rod is made from a material whose creep rate equation follows a power law relationship. It is used to connect two rigid plates inside a boiler that operates continuously at 400 °C. At this temperature, the material has the following properties: Young's modulus = 150 GPa, exponent in the power law equation, n = 7.5, while the coefficient, B = 5.2 x 10⁻²⁰ when stress is in MPa and the strain rate is expressed in % per hour. The initial force in the rod is 13 kN. Calculate the stress in the rod after four years of continuous service.

Answer:
The initial stress in the rod is:

$$\sigma_i = \frac{4P}{\pi d^2} = \frac{4 \times 13000}{\pi 0.02^2} = 41.4 MPa$$

The elapsed time in hours, t = 4 x 365 x 24 hours = 35 040 hours.
Substituting for initial stress and t in equation 11.13:

$\sigma = 22$ MPa.

## 11.8    PREDICTION OF LONG TERM PROPERTIES

Creep tests take very long to perform such that generation of creep data for use in design would be impractical. To overcome this, ways have been sought to enable designers to predict long-term properties by extrapolation of data collected over shorter periods. The short-term tests are performed at higher temperatures than the expected service temperature. Defining certain parameters that incorporate both temperature and time does this. One of the more popular ones is the Larson-Miller parameter $P_1$, defined as:

$$P_1 = T(\ln t + C_1) = f(\sigma) \qquad\qquad (11.14)$$

where T is the temperature on the absolute scale, $C_1$, the Larson-Miller constant lies between 35 and 60 and t is the time in hours. Tests are performed at many values of stress, temperature and times and the results used to plot a Larson-Miller plot or master curve. Such a plot is shown schematically in figure 11.8 and shows $P_1$ as a function of the stress.

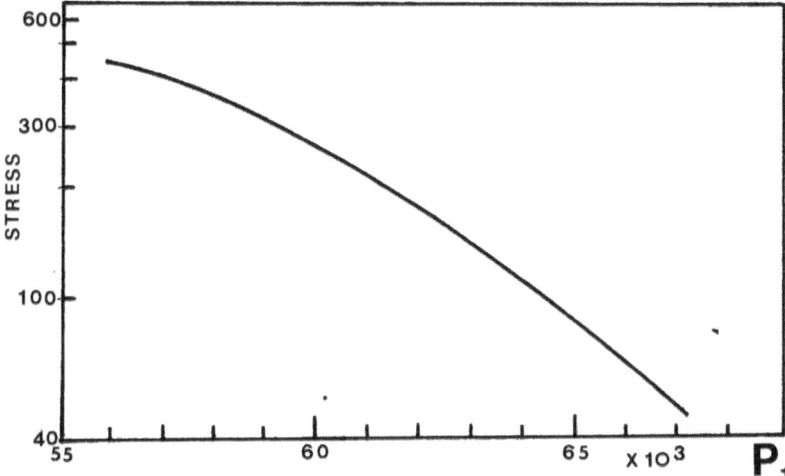

**Fig. 11.8** Schematic representation of a Larson--Miller plot

To use the master curve, the values of temperature and time desired for the design are used to calculate $P_1$. The master curve is then used to read the value of stress corresponding to the calculated value of $P_1$. This stress is used as the limiting stress in the design.

Several other parameters have been suggested including the Sherby-Dorn parameter, Manson-Haferd parameter among others. All are used in a similar manner. Care has to be taken though to ensure that no microstructural changes take place between the temperature of test and the temperature of service and that the creep deformation mechanism is the same at the two temperatures being traded.

## PROBLEMS

11.1 A gas turbine at Kipevu Power Station has blades, which are made from the material whose creep properties at 800 °C are given in figure Q11.1. The stress due to the centrifugal force at a radius r is given by:

$$\sigma = 1/2 \, \rho\omega^2[\, r_o^2 - r^2]$$

where $\rho$ is the density of the material, $\omega$ is the angular velocity in radians/sec., and $r_o$ is the tip radius. Given the following data, calculate the safe speed, in rpm, at which the turbine may operate: $\rho = 8500$ Kg/m³, $r_o =$ 200 mm, root radius = 140 mm, operating temperature = 800°C. Note: the strain should be limited to 1 % after $10^5$ hours. (Ans: 10,300 rpm).

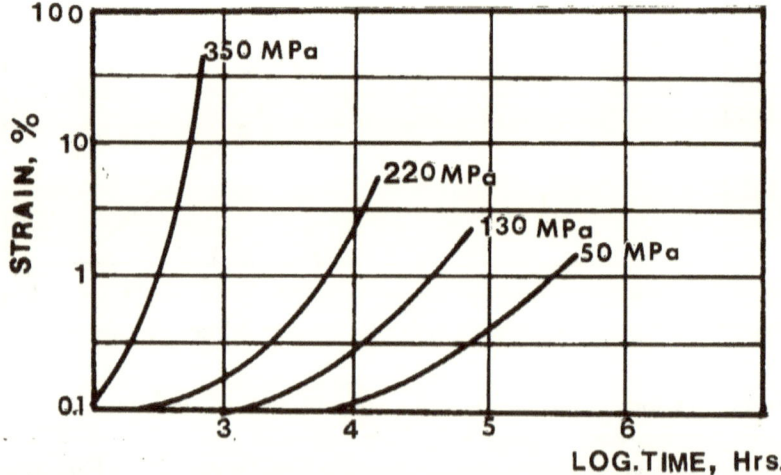

**Fig. Q11.1**

11.2 (a) List the mechanism responsible for low and high temperature creep.

(b) A material creeps according to the formula $d^\varepsilon/dt = B\sigma^n$ where $d^\varepsilon/dt$ is the creep rate, $\sigma$ is the applied stress and B and n are constants. It is established that at 430 °C, the modulus of elasticity E = 155 GPa and n = 8.0. A bolt of the material is tightened to a stress of 70 MPa and rigidly attached at both ends. The creep rate is found to be 3 x $10^{-8}$ %/h. What is the stress in the bolt after 3 years in service? Derive any formula used. (Ans: 48.4 MPa.)

11.3 (a) Give six reasons why creep rate increases with increase in temperature.

(b) What is "homologous temperature" and why is it important for creep studies?

11.4 The tubes of a steam generator have an outside diameter of 60 mm. The generator operates at a temperature of 400 °C and a pressure of 9.8 MPa. The elongation of the tube is to be limited to 1 % after 10,000 hrs, and a safety factor of 2 is to be used in the design. If the creep strength of the material of the tube at 400 °C, $\sigma_{1/10,000} = 185$ MPa, determine the required wall thickness. (Ans: 3.2 mm.)

11.5 Derive the formula for stress relaxation of a material fixed at both ends and creeping according to the law $\varepsilon = B\sigma^n$.

For a certain material, E = 150 GPa, n = 8.0, B = 5.2 x $10^{-23}$ % /MPa.h. A bolt of the material is tightened to 210 MPa. What is the stress in the bolt after 4 years of continuous service? (Ans: 47 MPa.)

11.6 Describe the creep behaviour of a metal and explain how it varies with stress.

11.7 Creep tests were performed on a high temperature alloy and the following data were recorded at 800 °C:

| Stress [MPa] | 67 | 89 | 119 | 168 | 237 |
|---|---|---|---|---|---|
| Creep rate [% /h] | 5.62 x $10^{-5}$ | 7.5 x $10^{-4}$ | 0.01 | 0.178 | 4.47 |

The material creeps according to the law $\varepsilon = D\sigma^m$ where D and m are constants. A rod made from the material is subjected to a load of 36 KN at 800 °C. Using a safety factor of 1.8, determine the necessary diameter based on an allowable strain of 1.5 % in 10,000 hrs. (Ans: 33 mm.)

11.8 The tubes of a steam generator have an outer diameter of 57 mm. The generator operates at 380 °C and 8.5 MPa. Determine the minimum wall thickness required if the properties of the material vary with temperature as below. A factor of safety of 1.5 is required against both creep and yielding. (Ans: 3.1 mm.)

| Temp. °C | 300 | 350 | 400 | 450 |
|---|---|---|---|---|
| $\sigma_{0.2}$ [MPa] | 160 | 140 | 130 | 110 |
| $\sigma_{1/10,000}$ [MPa] | - | 150 | 97 | 50 |

## Further Reading

Kraus, H., <u>Creep Analysis</u>, John Wiley, New York, 1980.

Pascoe, K.J. <u>An Introduction to the Properties of Engineering Materials.</u> 3rd ed. ELBS & Van Nostrand Reinhold (UK), Workingham, 1982.

Hertberg, R.W. <u>Deformation and Fracture Mechanics of Engineering Materials.</u> John Wiley, New York, 1977.

# CHAPTER TWELVE

# CLASSIFICATION OF FRACTURES

## 12.1  INTRODUCTION

The term fracture refers to the fragmentation or separation of a solid into two or more parts, under the action of a stress. It occurs in characteristic ways depending on the state of stress, the strain rate, the temperature and the properties of the material.

Broadly, fractures can be classified into two categories: brittle fracture and ductile fracture. These may be identified either by appearance or by the crystallographic mode. Ductile fracture occurs by shear and the resulting fracture surfaces look dull and fibrous. Brittle fracture surfaces have a crystalline and bright (granular) appearance and the fracture occurs by cleavage. The ease or difficulty with which fracture takes place is determined by the material's toughness. This is a measure of the amount of energy the material absorbs in fracturing. Tough materials undergo considerable plastic deformation before fracture thereby absorbing more energy than brittle materials, which undergo negligible plastic deformation.

The material properties percent elongation and percent reduction in cross section area may be used to estimate the toughness but in a few cases, materials showing high values of these properties may have low impact toughness. This is especially so for a material that is sensitive to rate of loading.

## 12.2  DUCTILE FRACTURE

In ductile fracture, the fracturing process is accompanied by considerable plastic deformation and gross deformation of the fractured surfaces. The three main forms of ductile fracture: shearing fracture, rupture and formation of cup and cone are illustrated schematically in figure 12.1. Of these, the most important for polycrystalline materials is the formation of cup and cone, which proceeds (in slow loading situations) as follows:

(a) Necking begins when the increase in strength due to strain hardening cannot compensate for the decrease in cross section area. This occurs at the maximum load.

(b) The necking leads to a triaxial state of stress (at the neck), the hydrostatic component of which acts along the axis at the centre.

(c) Fine cavities or voids develop at inclusions, second phase particles, and grain boundary triple points in the material. The voids result in even further reduction in cross section area and hence an increase in stress.

(d) The voids coalesce into a crack perpendicular to the direction of the principal stress.

(e) Near the surface, the crack propagates along localized shear planes to form the cone.

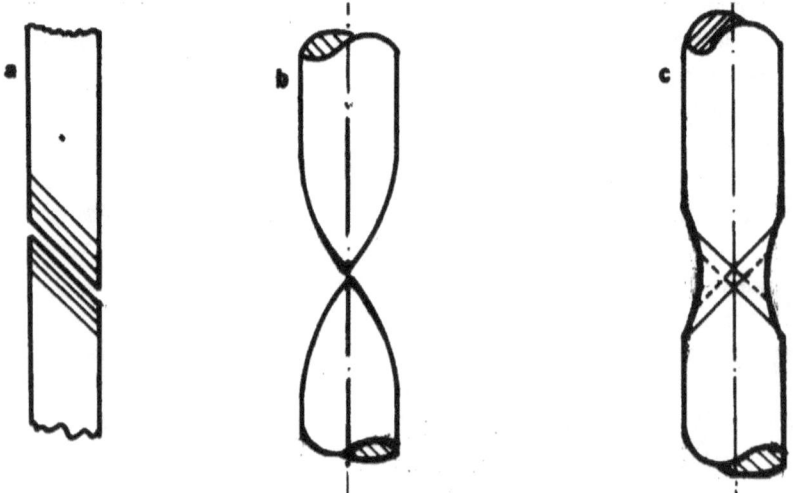

**Fig. 12.1** Forms of ductile fracture (a) shear (b) rupture (c) formation of cup and cone

These steps are depicted schematically in figure 12.2. In other words, ductile fracture occurs by a process of void coalescence. Due to the voids, the fracture surface consists of dimples leading to the dull fibrous appearance.

## 12.3  BRITTLE FRACTURE

The type fracture, which occurs without appreciable plastic deformation is termed brittle fracture. It occurs by rapid crack propagation, without any prior warning signs (e.g., excessive strain, necking, etc.) and proceeds at very high speed through the material. This type of failure therefore has to be avoided at all costs. The tendency to brittle fracture in materials is increased by decreasing temperature, increasing strain rate and by the presence of a triaxial state of stress such as that produced by notches and other discontinuities. Typical stress-strain curves expected from brittle and ductile materials are shown in figure 12.3. In order to control the occurrence of brittle fracture in materials, it is necessary to characterize the influence of cracks on the strength of a material and in particular to identify the conditions under which fast propagation of a crack takes place when stress is applied to a cracked component.

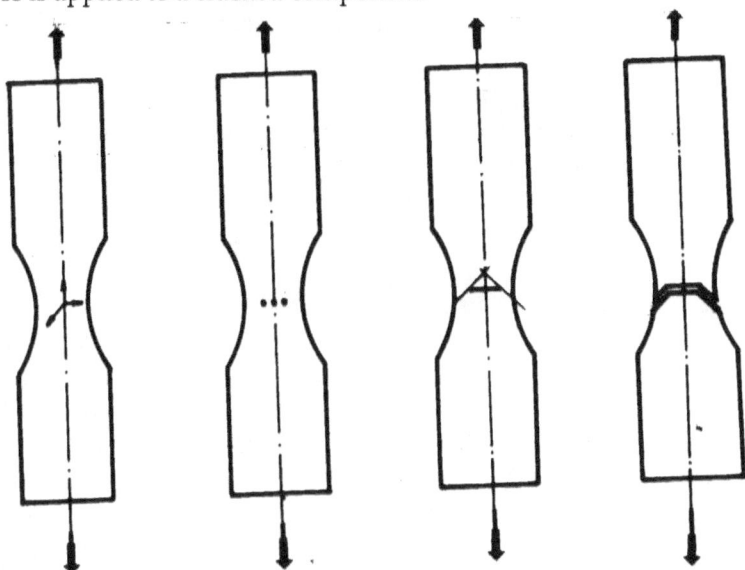

**Fig. 12.2** Stages in the formation of cup and cone

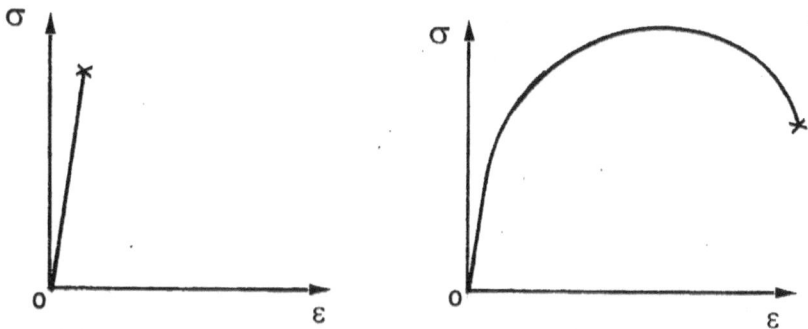

**Fig. 12.3** Stress-strain curves for (a) a brittle material (b) a ductile material

## 12.3.1 Theoretical Strength of Materials

Brittle fracture occurs when the separation of atoms of the material occurs by cleavage i.e., by the atoms literally pulling apart. The theoretical value of stress required to cause this separation can be shown to be equal to one tenth of the Young's modulus of the material (see references at the end of the chapter). In practice, the cleavage strength of most materials is only about one hundredth of their Young's modulus. The reason for this discrepancy is the presence in real materials of crack like defects. The need to characterize the effect of cracks in members becomes apparent.

## 12.3.2   The Griffith Theory

This was the earliest attempt to characterize conditions under which unstable crack extension takes place. Here, the energy balance in a stressed cracked material is considered. To determine the conditions under which crack extension takes place, two forms of energy are considered. First, the elastic strain energy: Any strained elastic member stores the work done in straining it in the form of strain energy. If a crack is now introduced in the elastic member, the stiffness is reduced. As a result, the stored elastic strain energy is reduced (the balance is released). However, the introduction of the crack creates new surfaces, and energy is required for this. Griffith reasoned that unstable crack extension (brittle fracture) takes place if an increase of crack length resulted in more strain energy being released than that required to create new surfaces. For example in an infinite plate loaded with an infinitely applied stress, σ (figure 12.4) having a crack of length 2a, introduction of the crack results in a <u>release</u> of elastic strain energy, $U_e$, given (for plane stress) by:

$$U_e = \frac{\pi \sigma^2 a^2 B}{E} \qquad (12.1)$$

For plane strain, the expression becomes:

$$U_e = \frac{(1 - v^2)}{E} \pi \sigma^2 a^2 B \qquad (12.2)$$

In the above equations, B is the thickness; E the modulus of elasticity and v is Poisson's ratio. (Equations 12.1 and 12.2 have been stated here without proof. The reader interested in their proof is referred to any textbook on fracture mechanics.)

The surface energy, $U_s$, resulting from the creation of crack surfaces is:

$$U_s = 4\gamma_s a B \qquad (12.3)$$

194

where $\gamma_s$ is the specific surface energy per unit area and the factor 4a comes due to the fact that in creating a crack of length 2a, the area of the created surfaces = 2a x 2 = 4a per unit thickness. Thus for plane stress, the total energy, U after introduction of the crack is given by:

$$U = U_o + U_s - U_e \qquad (12.4)$$

$$U = U_o + 4\gamma_s aB - \frac{\pi \sigma^2 a^2 B}{E} \qquad (12.5)$$

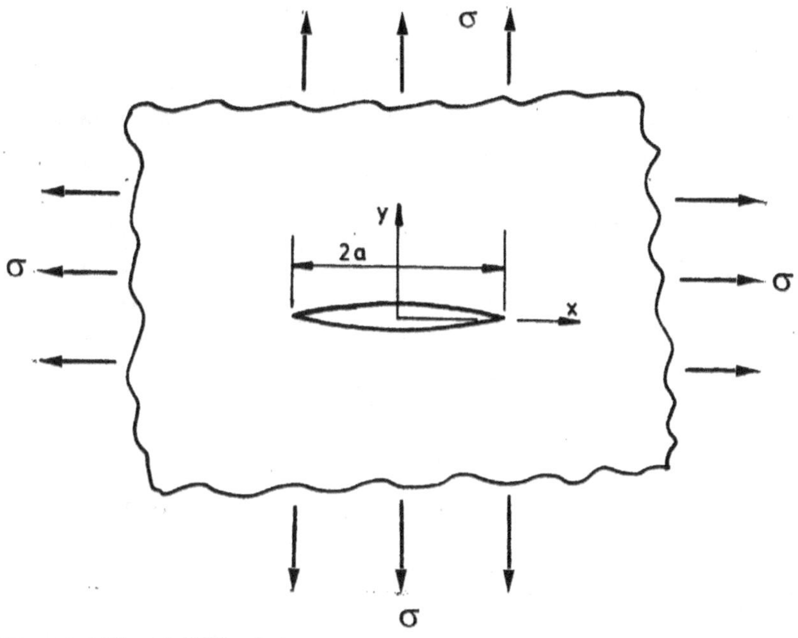

**Fig. 12.4** The Griffith plate

Consider now the effect of extending the crack. Let the crack extend by an amount, da. The resulting change in energy can be expressed (for plane stress) as:

$$\frac{dU}{da} = 4\gamma_s B - \frac{2B\pi \sigma^2 a}{E} \qquad (12.6)$$

At the point of instability, $dU/da = 0$ from which:

$$(\sigma^2 a)_{cr} = \frac{2\gamma_s E}{\pi} \qquad (12.7)$$

where $(\sigma^2 a)_{cr}$ is the critical combination of stress and crack length required to initiate unstable crack growth, since the right hand side of equation 12.7 is a constant for any given material. A similar equation can be developed for plane strain by replacing E with $E/(1 - v^2)$.

Experiments carried out by Griffith on glass, which is nearly an ideally brittle material, confirmed the validity of his argument for such materials. For other materials however, some plastic deformation occurs at the crack tip absorbing most of the released strain energy. Moreover, the development of equation 12.7 is based on the theory of elasticity and assumes an ideally sharp crack. Furthermore, the approach allows only two states: a stable non-growing crack, or a crack growing in an unstable manner. In practice, some cracks, e.g., fatigue cracks, are known to grow in a slow stable manner. Griffith's theory in the form given in equation 12.7 is therefore inapplicable to engineering materials without modification. To take account of the plastic deformation for example, the Griffith equation was modified by Irwin and Orowan independently to:

$$(\sigma^2 a)_{cr} = \frac{2E(\gamma_s + \gamma_p)}{\pi} \approx \frac{2E\gamma_p}{\pi} \qquad (12.8)$$

where $\gamma_p$ (which is usually several orders of magnitude larger than $\gamma_s$) is the plastic work factor, which accounts for the energy absorbed in plastically deforming the crack tip. Equation 12.7 is referred to as the modified Griffith equation.

The Griffith equation outlined above forms the basis of fracture mechanics, which is the modern tool used in the analysis of the strength of all structures and components containing cracks or involving the growth of cracks.

## 12.4   BRITTLE-DUCTILE TRANSITION
### 12.4.1 Notched Bar Impact Tests
The toughness (or more correctly the notch toughness) of materials is determined by notched bar impact tests. The most common of these are the Charpy-Vee-Notch (CVN) tests and the Izod tests (see KS-06-902: Kenya Standard Specification for Charpy Impact (V-Notch) for Steels). In both

tests (CVN and Izod), a pendulum with a known mass, m, falls a known height, $h_1$, and hits and breaks a prepared notched specimen, and then rises to a new height, $h_2$. The energy used in breaking, or tearing the specimen, designated $C_v$, is then:

$$C_v = mg(h_1 - h_2)$$
(12.9)

where g is the acceleration due to gravity. By observing the fractured surfaces, the percentage of brittle fracture shown on the surface can also be estimated.

The CVN and Izod tests differ in the details of preparation of the specimen. The two types of specimens are shown in figure 12.5. The impact load is also applied differently: In the CVN, the specimen, which is a 10 mm x 10 mm square bar, is held as a simply supported beam and loaded at the centre while in the Izod, the specimen (a 10 mm diameter rod) is loaded in bending but as a cantilever.

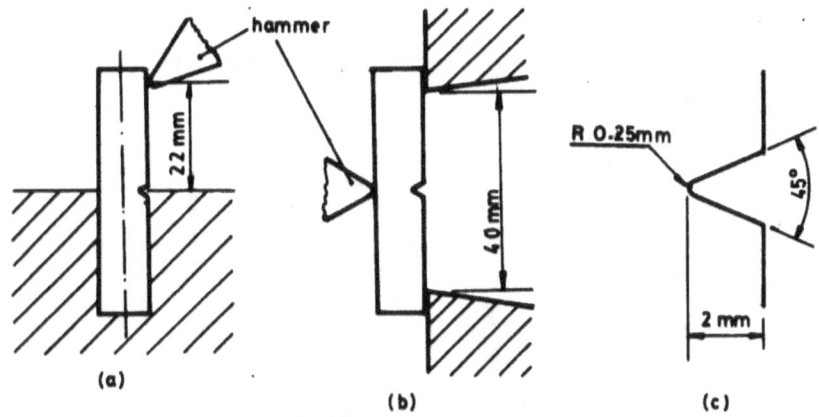

**Fig. 12.5** Impact test specimens: (a) Izod (b) CVN (c) details of notch

There are 3 important differences between notched bar impact tests and tensile tests:

   (i) In the impact tests, the specimen is notched. This leads to a triaxial state of stress at the root of the notch, when the load is applied. In tensile tests, a uniaxial state of stress prevails in the gauge length portion of the specimen.

   (ii) The strain rate in impact tests is much higher (approximately $10^3/s$) when compared with the tensile test.

(iii) The load in impact tests is a bending load as opposed to that in a tensile test that is tensile.

Considerable scatter is observed in the results from notched bar impact tests mainly due to:

(a) Local variations of material properties. In notched bar tests, the result recorded is the toughness of the particular place at which the notch is cut. Since local properties of the material may vary from point to point, scatter in the results is expected.

(b) Non-reproducibility of notches. No measurement of the actual notch dimensions is done in notched bar impact tests and since there may be slight variations in the notch dimensions, a scatter in the results is noted.

(c) Placement of specimen. Any deviation from the optimum position of application of the blow greatly affects the state of stress at the notch tip.

## 12.4.2 Transition Temperature Curve

When the notch toughness of various materials is determined over a range of temperatures, the variations shown in figure 12.6 result.

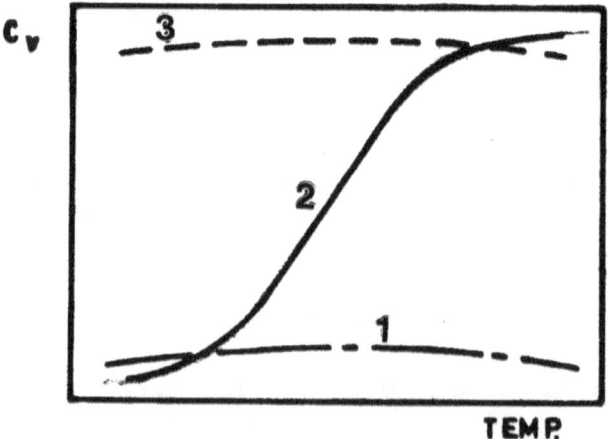

**Fig. 12.6** Variation of impact energy with temperature (1) high strength BCC alloys (2) low strength BCC alloys (3) FCC metals and alloys.

In general, FCC metals and alloys show a relatively high toughness at all temperatures. The main reason is that, in these materials, there are many slip systems making cleavage almost impossible. High strength BCC alloys have low toughness at all temperatures. Low strength BCC alloys (which include steel), zinc, beryllium, and ceramics show a transition temperature phenomenon. That is, they are brittle at low temperatures but as the temperature is raised, they show a transition to ductile behavior. It is however, noticed that the transition takes place over a range of temperatures (and not at a distinct temperature). It is therefore not possible to define a unique transition temperature and several definitions are used (see figure 12.7):

$T_1$: The temperature at which the fracture surface shows a 100 % fibrous (ductile) appearance. This is termed the fracture transition plastic (FTP).

$T_2$: The fracture appearance transition temperature (FATT). The temperature at which the appearance of the fracture surface shows 50 % brittle and 50 % ductile appearance.

$T_3$: The temperature at which the absorbed energy $= 1/2 (E_{max} + E_{min})$.

$T_4$: The ductility transition temperature (DTT). The temperature at which $C_v$ has a given defined value.

$T_5$: The nil ductility temperature (NDT). The temperature at which fracture surface shows 100 % brittle appearance.

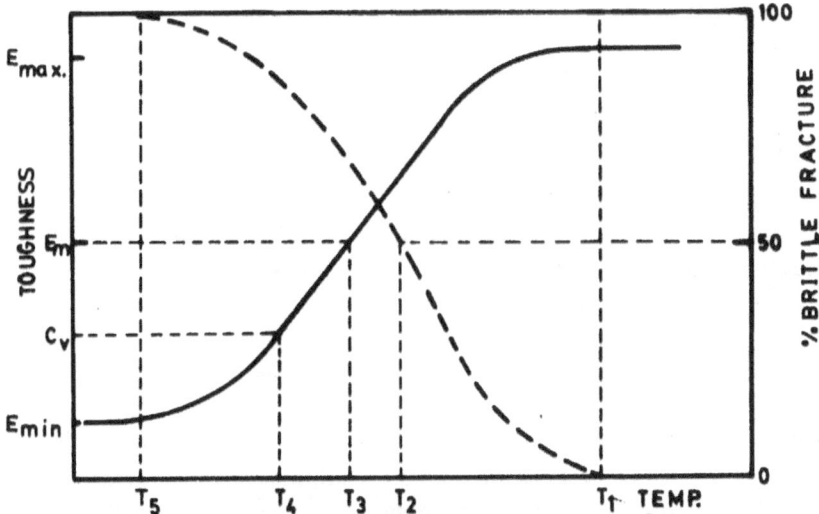

**Fig. 12.7** Transition temperature curve: variation of impact energy (solid line); variation of % brittle fracture (dotted line)

In general, the transition temperature is 0.1 to 0.2 of the melting temperature, $T_m$ on absolute scale for metals and 0.5 to 0.7 $T_m$ for ceramics.

199

The results of notched bar impact tests cannot be used directly in design because the magnitude of the triaxial state of stress at the notch root is not known. Furthermore, there is no way to take the size and geometry of the notch, nor the strain rate, into consideration. The results are used indirectly in design in what is known as the "transition temperature approach" to design. The impact tests are carried out to select a material with the required toughness at a given temperature and then another criterion e.g., endurance limit or tensile strength is used for design. Thus the method is used to select a material out of many candidate materials and to assess the suitability of various heat treatments for avoiding low temperature brittleness.

The disadvantages of the transition temperature approach mentioned above led to the search for a better method for the control of brittle fracture. The new method is the fracture mechanics approach mentioned in section 12.3.2. In this method, a stress intensity factor, K, (which incorporates both the stress and the crack length) is defined. A crack will propagate in an unstable manner if the material is loaded such that the stress intensify factor applied exceeds the materials fracture toughness, $K_{Ic}$. The subscript I indicate the mode of loading, while c stands for critical. $K_{Ic}$ is a material property that can be determined in the laboratory. The design then involves limiting the applied stress intensity to a fraction of $K_{Ic}$. More about fracture mechanics cannot be said here and interested reader is referred to textbooks on fracture mechanics.

## 12.4.3 Factors Affecting Transition Temperature and Toughness in Steels

The transition temperature in steels is affected by several factors including the chemical composition, impurities, environment, etc.

(a) Chemical Composition

An increase in the carbon content lowers the $C_v$ at any given temperature and raises the transition temperature as shown schematically in figure 12.8. It may be observed from figure 12.8 that the transition temperature becomes better defined at lower carbon contents. However, it should be kept in mind that lower carbon content result in steel with lower strength.

An increase in the content of manganese is beneficial in that it lowers the transition temperature and raises the strength. Unfortunately, increased content of manganese also lowers the weldability of the steel. To get the best compromise, the manganese to carbon ratio is kept at approximately 3:1.

The other alloying elements that lower the transition temperature include nickel and vanadium while silicon, phosphorous and molybdenum raise it. Chromium has little effect on the transition temperature.

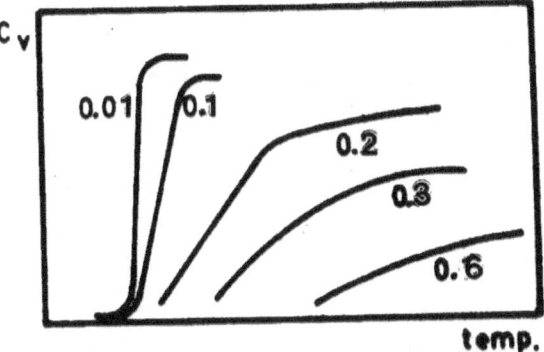

**Fig. 12.8** A schematic showing variation of transition temperatures for normalized plain carbon steels with different carbon contents.

<u>(b) Gaseous Impurities and Environment</u>
The presence of gaseous impurities (oxygen, nitrogen and hydrogen) in steels has a very adverse effect on the transition temperature. Oxygen is the worst. Concentrations as low as 0.06 % raise the transition temperature by about 400 °C. This effect is countered in steel making by "killing" (or de-oxidation).

Hydrogen may be introduced into steel either during melting and casting or during manufacturing processes like welding. It raises the transition temperature in concentrations as low as 0.0001 % (or prevents transition to ductile behavior altogether) by a process termed hydrogen embrittlement. The embrittlement is made worse by the fact that it is most effective at mid-temperatures (which includes room temperature) and at slow strain rates. It is thought that a hydrogen embrittlement result from a pressure build up inside the lattice after the hydrogen has broken down to hydrogen atoms but more on its mechanism is beyond the scope of this book.

Nitrogen also raises the transition temperature but its effects are less dramatic compared to those of hydrogen and oxygen.

Most steels are sensitive to particular environments. These environments render the steels brittle and raise their transition temperature. This effect, which is termed stress corrosion cracking (i.e., the combined effect of a static stress and a corrosive environment), is very particular to an alloy/environment combination and involves fairly complex mechanisms.

(c) Size (Thickness)

An increase in the thickness of a member leads to a decrease of toughness due to the increased tendency of thick specimens to plane strain. Another contributing factor is the fact that thicker specimen tend to have coarser grains. The effect levels off after a certain thickness when full plane strain conditions have been achieved.

(d) Others

Other less important factors include grain size (a coarser grained structure has less toughness); orientation (a steel will show higher toughness when the crack grows perpendicular to the orientation of the grains); cold working (which decreases the toughness) and tempering temperature.

## 12.5   NOTCHES IN BRITTLE FRACTURE

It will be recalled from the study of solid mechanics (strength of materials) that notches in statically loaded members lead to stress concentration. In impact loading, notches lead to a triaxial state of stress and hence increase the tendency to brittle fracture by the following steps:

(a) The stress concentration at the root of the notch leads to local yielding at the root.

(b) Due to Poisson's effect, yielding in one direction leads to an attempt of the material to contract in the other two directions.

(c) But, the material around the root of the notch is still elastic and will resist this attempted contraction hence stresses will be set up in these directions. Thus a tri-axial state of stress will result from the uniaxial loading.

(d) With a triaxial state of stress, the yield criterion changes from the simple case of uniaxial stress to a more complex situation. In effect the material does not yield until a higher applied stress is attained i.e., the yield stress is increased.

(e) A ductile fracture starts at the notch.

(f) The crack leads to even further stress concentration (intensification).

(g) Local straining at the crack tip leads to local strain hardening thereby increasing the yield stress further.

(h) The above factors combine and lead to local stresses higher than the stresses required to overcome cohesive forces and the crack extends in a brittle manner.

Likewise, factors tending to raise the yield stress like low temperature and high strain rates increase the tendency towards brittle fracture.

## PROBLEMS

12.1 The presence of a notch in a component leads to a tri-axial state of stress and hence increases the tendency towards brittle fracture. Explain how this happens.

12.2 Using the "transition temperature diagram", give five different points that are used to define transition temperature in fracture.

12.3 A high strength steel has the following properties: E = 200 GPa, $\sigma_{UTS}$ = 1950 MPa, specific surface energy = 2 J/m², plastic work factor = 2 x 10³ Jm⁻². Calculate the percentage reduction on strength caused by the presence of a 4 mm long centre crack perpendicular to the direction of stress in a large plate of the material. Assume plane stress conditions. (Ans: 42 %.).

12.4 Give 3 methods that are used for impact testing of full-scale structures. Sketch schematically, on the same diagram, the variation of fracture energy with temperature for (i) low strength BCC metals/alloys (ii) high strength BCC alloys (iii) FCC metals/alloys.

12.5 With the aid of sketches, explain the process of formation of cup and cone in ductile fracture.

12.6 How would you distinguish between brittle and ductile fracture surfaces by (i) crystallographic mode (ii) appearance?

12.7 With the aid of sketches, illustrate 3 forms of ductile fracture that may be encountered in various materials.

12.8 Write notes on six major factors that affect the transition temperature in steels.

12.9 (a) What are the major differences between tensile tests and notched bar impact tests?
(b) With the aid of a sketch, explain what is meant by (i) Fracture

Transition Plastic (FTP), (ii) Nil Ductility Temperature (NDT), (iii) Fracture Appearance Transition Temperature (FATT), (iv) Ductility Transition Temperature (DTT).

12.10 A 40 mm thick high strength steel plate is to be used for fabrication of an oil tanker. The tanker will be used in conditions where the temperatures may reach -40 °C. Knowing that after construction, the tanker is tested by a non-destructive method that will reveal any cracks longer than 2 mm. calculate the maximum stress that may be applied to the plate without causing brittle fracture. The properties of the material at -40 °C are: $v = 0.3$; $\sigma_{UTS} = 1600$ MPa; E = 250 GPa; $\gamma_s = 4$ J/m$^2$; $\gamma_p = 3000$ J/m$^2$. A factor of safety of 1.5 is required. Assume plane strain. (Ans: 483 MPa.)

12.11 A large steel plate, 60 mm thick, is made from a special steel with the following properties at -60°C: $v = 0.28$; $\sigma_{UTS} = 1800$ MPa; E = 260 GPa; $\gamma_s = 5$ J/m$^2$; $\gamma_p = 3300$ J/m$^2$.
Due to the method of fabrication, cracks up to 1.5 mm may be present in the plate. Calculate the maximum stress to which the plate may be subjected if a safety factor of 1.8 is desired against brittle fracture. You may assume plane strain conditions. (Ans. 494 MPa.)

12.12 Using a sketch of the variation of toughness with temperature for steels, show and define five different temperatures that are used to define the transition temperature from brittle to ductile fracture.

## Further Reading

Dieter, G. Mechanical Metallurgy, McGraw-Hill, New York, 1986.

Hertberg, R.W. Deformation and Fracture Mechanics of Engineering Materials, John Wiley, New York, 1977.

Kavishe, F. Fracture Mechanics, Jomo Kenyatta Foundation, Nairobi, 1998.

# CHAPTER THIRTEEN

# INTRODUCTION TO POLYMERS

## 13.1   INTRODUCTION

The word polymer is a Latin word meaning "many-membered". It is used to describe materials made up of long chain macromolecules in contrast to, say, metals where the units making up the material are single atoms. Polymers may be natural or synthetic. The natural polymers include natural rubber ((poly)isoprene); protein (the basis of all animal life including flesh, bone, skin and cartilage); and cellulose and lignin (the main materials in the cell structure of plants and hence the basis of materials like wood, jute, cotton, sisal, and hemp). In addition, there are several man made products based on natural polymers. Thus leather, silk and wool are protein based while cork and felt are based on cellulose and lignin. In synthetic polymers, the macromolecules are chemically synthesized through a polymerization process. Synthetic polymers have become common place articles in everyday use. Polyvinyl chloride (PVC) articles, foam for cushions and mattresses, are common examples.

Polymers may also be classified as being either organic or mineral (inorganic). In organic polymers, carbon atoms form the backbone of the macromolecule while in mineral polymers; the backbone may be silicon in the form of silicon dioxide (e.g., silicone rubber) or alumina, $Al_2O_3$, or a combination of both.

Natural polymers and their derivatives (e.g., wood, silk, cotton, wool, paper, rubber, etc.,) have been used by man for many centuries. The use of synthetic polymers has however, accelerated over the last few decades (and the growth continues) in three main areas of industry:

(1) Textiles industry where use of artificial fibers like polyester and nylon (polyamide) has greatly increased.

(2) Structural and mechanical design where polymers continue to supplement metals and other traditional materials.

(3) In the fields of photo electronics (liquid crystal display devices) and computers.

The following properties helped accelerate the use of organic polymers for engineering purposes:

(a)     Polymers have low melting point and hence are easier to fabricate.

b)     They have low specific gravity (typically 0.9 to 1.5) and hence are particularly useful where low weight is an important consideration.

c) They have a wide range of properties.
d) They are resistant to atmospheric corrosion and hence do not need to be oiled or coated to prevent corrosion.
e) They have sound and vibration absorbing properties.
f) They have a low coefficient of friction
g) They have good electricity and heat insulating properties.
h) They have found great use as matrix materials in manufacture of composite materials.

Despite these advantages organic polymers suffer from some serious disadvantages of which the designer intending to use them must be aware:

(a) They are very sensitive to temperature. Their properties decrease significantly with small increases of temperature and there is possibility of chemical decomposition.
(b) They exhibit, even at room temperature, time dependent variation in mechanical properties.
(c) Most of them undergo environmental degradation mainly due to sunlight (more exactly ultra-violet radiation) and exhibit an aging phenomenon.
(d) Some of them cannot be recycled leading to problems of waste disposal.
(e) The synthetic polymers (plastics) are based on petroleum and hence are affected by the supply/pricing etc., of petroleum products.
(f) Most polymers have high coefficient of thermal expansion leading to problems of thermal stresses.

## 13.2 THE STRUCTURE OF POLYMERS

In order to be able to use polymers effectively, one must understand their structure. In the rest of the chapter, we will concentrate on artificial organic polymers (also called plastics) since these are of most importance to the engineer.

### 13.2.1 The Basis of Polymer Structure (Polymerization)

Polymers are made up by joining "monomers" into long chains in a chemical process known as polymerization. Monomers contain unsaturated bonds. These are double bond between two carbon atoms. The double bonds can be broken thereby allowing the molecule (now called "mer") to join itself to other mers. There are two types of polymerization: addition and condensation polymerization.

Addition polymerization involves the use of an initiator or catalyst, pressure and/or heat. The initiator e.g., hydrogen peroxide, starts the reaction by providing very reactive -OH radicals which attach themselves to one of the bonds in the double bond. This leaves an activated monomer, which reacts with the nearest molecule. This process continues until either:

(a)  A terminator (e.g. an-OH group) attaches itself at the end of the chain thus stopping further reaction.

(b)  The chain links with another growing chain.

(c)  The monomers are finished.

The simplest example of this is the polymerization of polyethylene:

$$n\left[\begin{array}{cc} H & H \\ | & | \\ C = C \\ | & | \\ H & H \end{array}\right] \implies \left[\begin{array}{cc} H & H \\ | & | \\ -C--C- \\ | & | \\ H & H \end{array}\right]_n \tag{13.1}$$

In condensation (or step growth) polymerization, two different molecules react to produce a larger molecule and a small molecule (usually water) as a by-product. If the molecules are bifunctional i.e., if they have only one double bond, no polymer will result. But if one or both of the molecules is polyfunctional, the remaining bond will join with other molecules to form the polymer, e.g., a double organic acid and a double alcohol reacting to give polyester and water (see equation 13.2).

$$n.R_1 \begin{array}{c} CH_2OH \\ | \\ | \\ CH_2OH \end{array} + \quad n.R_2 \begin{array}{c} COOH \\ | \\ | \\ COOH \end{array} \implies$$

Alcohol    Acid

$$nHO\text{-}[CH_2\text{-}R_1\text{-}CH_2\text{-}O\text{-}C\text{-}R_2\text{-}C\text{-}O]_n \ + \ nH_2O \tag{13.2}$$
$$\underset{||}{\phantom{nHO}} O$$

Polyester                    Water

It should be clear from above that in bulk polymers, different chains will have different lengths (number of monomer units). The number of monomer units that make up a chain is termed the degree of polymerization (DP). The physical and mechanical properties of polymers depend on their DP. The molecular weight distribution depends on the method of manufacture as well as the additives, etc., which are added during manufacture. An average molecular weight, which is determined by measurement of physical properties like viscosity, can be defined in several ways. The number average molecular weight, $M_{n.av.}$ is obtained by dividing the chains into size ranges, $i = 1, 2, ......, k$, then finding the number of chains within each range. Then:

$$M_{n.av} = \Sigma \, x_i \, M_i \qquad\qquad (13.3)$$

where $x_i$ is the fraction of the total number of chains within a size range, $i$, while $M_i$ is the mean molecular weight of the same size range.

Alternatively, a weight average molecular weight, $M_{w.av}$, based on the weight fractions of molecules within the size ranges may be defined as:

$$M_{w.av} = \Sigma \, w_i \, M_i \qquad\qquad (13.4)$$

where $M_i$ has the same meaning as above, while $w_i$ is the weight fraction of the chains or molecules within the same size range. The degree of polymerization mentioned previously may also be expressed as a number average or a weight average. Thus:

$$n_n = \frac{M_{n.av}}{m} \qquad\qquad (13.5)$$

$$n_w = \frac{M_{w.av}}{m} \qquad\qquad (13.6)$$

where, $n_n$ and $n_w$ are the number and weight degrees of polymerization, respectively, $m$ is the molecular weight of the monomer unit, while the other terms are as defined above.

An index which indicates the degree of homogeneity of molecular weight distribution within the polymer is the dispersity, $D$, defined as:

$$D = \frac{M_{w.av}}{M_{n.av}} \qquad\qquad (13.7)$$

## 13.2.2 Some Common Plastics

It may be useful at this point to mention the most common general purpose engineering plastics and their areas of application. These are:

(a) Plastics based on the vinyl monomer

```
    H     H
    |     |
    C  == C
    |     |
    H     R₁
```

When $R_1 = H$ the result is polyethylene, PE, which has been considered above. It is used for plastic tubing, plastic bottles, plastic cups and for electrical insulation. When $R_1 = CH_3$ the resulting polymer is polypropylene, PP, whose uses as the same as those of PE. When $R_1 = Cl$ we get polyvinyl chloride, PVC. This is used for imitation leather, plastic hoses and tubing, and as insulation for electrical wires. When $R_1 =$ benzene ring, the result is polystyrene, PS. It is mostly used for packaging and mouldings because of its light weight (low density).

(b) Plastics based on vinilydene group. In this group, other side groups replace two hydrogens in the ethylene molecule. The most common plastic in this group is polymethylmethacrylate, PMMA (also called Perspex or Plexiglas), which has the mer:

```
    H     CH₃
    |     |
  - C -   C-
    |     |
    H     COOCH₃
```

It is used for transparent articles, glass for windows and in laminated windscreens due to its transparency.

(c) Polytetrafloroethylene, PTFE (Teflon). All the hydrogen atoms in ethylene are replaced by fluorine. Thus the mer is:

```
      F
      |
   -- C --
      |
      F
```

PTFE has good resistance to temperature and low friction coefficient. It is therefore used in pipe joining and in the manufacture of non--stick cooking pans.

(d) Nylon 66. The repeat unit is $--C_6H_{11}NO--$ and it is used to make synthetic fibers in the textile industry and for manufacture of ropes.

(e) More complex plastics. These include:

Epoxy:

$$\begin{array}{ccc} CH_3 & & OH \\ | & & | \\ -O--C_6H_4--C--C_6H_4--O--CH_2--CH--CH_2-- \\ | \\ CH_3 \end{array}$$

It is used in adhesives and as a matrix for composite materials.

Polyester:

This is used for making synthetic fibre in the manufacture of textiles like Terylene; as laminated plates and as a matrix for composite materials.

Phenol-formaldehyde:

$$\begin{array}{c} OH \\ | \\ --C_6H_2--CH_2-- \\ | \\ CH_3 \end{array}$$

Phenol-formaldehyde is used in manufacture of Bakelite, tufuol and Formica. Urea formaldehyde is used for making electric fittings while melamine formaldehyde is used for manufacture of plastic tableware.

## 13.2.3    Polymer Morphology and Architecture

Even though the polymer chain is drawn as a straight line, it has a zigzag structure. The reader will recall from organic chemistry that the bonds in carbon have an equilibrium angle of 109.5° between them. Thus a straightened polyethylene chain for example has the form shown in figure 13.1. Furthermore, there is rotation of the other bonds about the C-C bond, (in an attempt to achieve the lowest potential energy). Where many chains are present, this makes the chains highly coiled and entangled.

If a monomer has an asymmetric C-atom, the nature of bonding in different monomers is not unique. The polymers may exhibit tacticity that describes the stereo regularity of the chain. The possible arrangements are: (see figure 13.2).

   (i)   Isotactic. In this arrangement, all the side groups are situated on one side of the chain only.

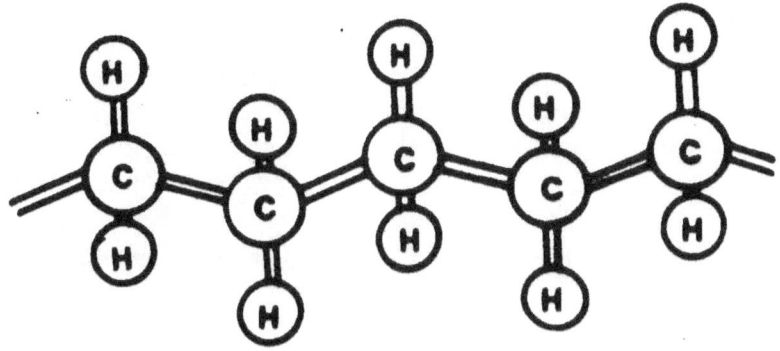

**Fig. 13.1** Shape of the polyethylene molecule

(ii)    Syndiotactic. Here, the side groups are attached on alternating sides of the chain as shown in figure 13.2 (b).

(iii)   Atactic. The arrangement of the side groups is random. Atactic polymers have low packing due to the difficulty involved in arranging the molecules in any sort of order. As a consequence, most of them are also amorphous.

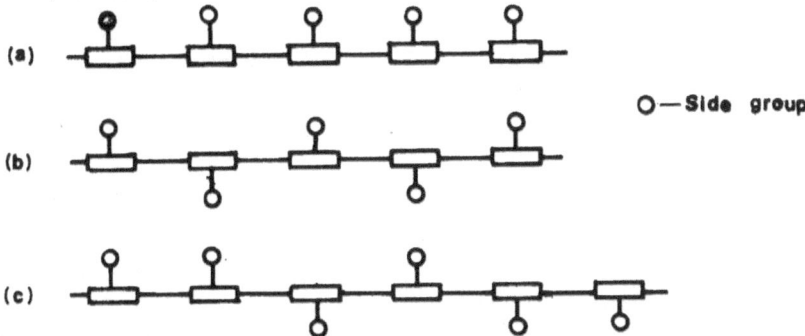

**Fig. 13.2** Tacticity in polymers (a) isotactic (b) sindotactic (c) atactic

Polymers such as polystyrene and PMMA may exist in any one of the three atacticities depending on the stereospecific catalyst used in their synthesis.

So far, it has been assumed that the whole polymer is made up of one type of monomer unit. This is true for most polymers. A polymer in which there is only one type of mer is termed a homopolymer. It may also be possible to have a polymer in which two or more types of monomer units make up a single chain. Such a polymer is termed a copolymer. There are statistical copolymers with regular or random arrangements of the monomers as shown in figure 13.3 and block copolymers in which different types of monomer units are arranged in blocks. Finally, the second monomer may be grafted onto the chain made up of monomer units of the first monomer. Examples of copolymers are polyvinyl chloride-acetate and rubber ((poly)isoprene):

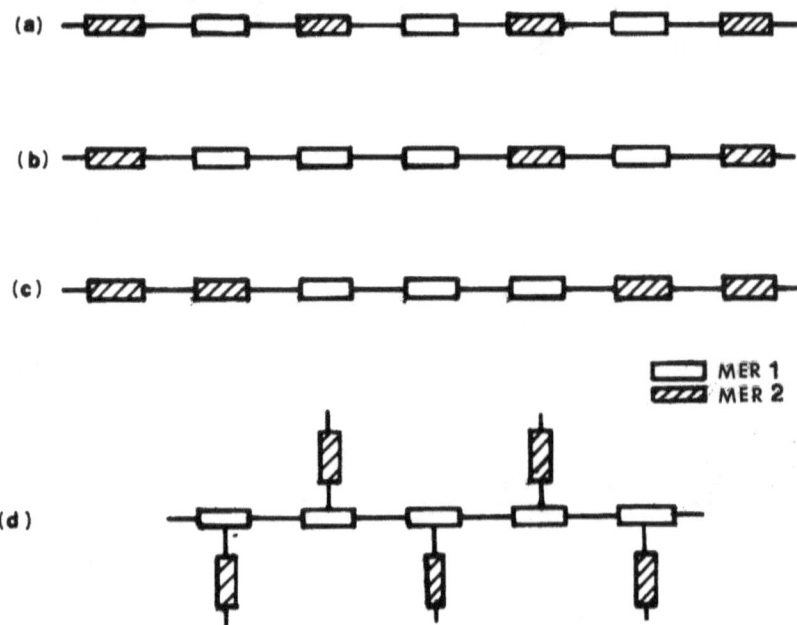

**Fig. 13.3** Copolymers (a) regularly positioned (b) randomly positioned (c) block copolymers (d) graft copolymer

## 13.2.4 Crystallinity in Polymers

Though polymers are in general amorphous, there are some that show at least partial crystallinity (ordered arrangement). The factors that determine the crystallinity are: whether the polymer is straight, branched or a network, regular or with side groups, the tacticity, etc. Regular, unbranched chain molecules form crystals easier. This is because such regular molecules are easier to arrange into some sort of order.

The ordered regions of polymers form by repeated chain folding into a lamella. The lamellae grow or stack along radial directions, trapping amorphous material between hence forming a spherulite as shown in figure 13.4. Each line in the figure represents a polymer chain. Polymer crystals may also form by twisting around the main chain to a helix like structure as in PTFE.

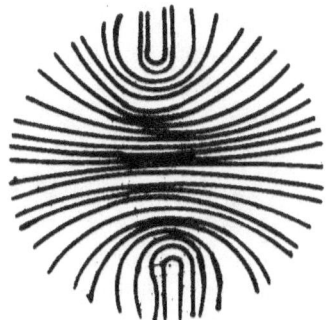

**Fig. 13.4** Schematic representation of a spherulite

There are several properties associated with crystallinity in polymers, crystalline polymers tend to be opaque or translucent e.g., PP, PA, PTFE, HDPE (high density polyethylene), nylon 66. They have higher density and higher strength. They have a fairly well defined melting point (the little variation present is due to differences in molecular weight) and do not exhibit the glass transition behaviour, which is expounded on in latter sections.

Polymers that are cross linked (see later sections), branched or which have bulky side groups do not form crystals easily. These amorphous polymers like PS; PVC, epoxy, elastomers, are, in general, transparent, have low density, and are isotropic.

### 13.2.5    Polymer Additives

Several other materials are added to polymers for specific purposes. Some of these are: stabilizers which are added to suppress breakdown of the polymer chains, fillers enhance certain properties and lower the cost e.g., carbon black which is added to rubber, plasticizers which increase the processability of the polymer, pigments and dyestuffs which give colour to the product. The exact chemical natures of most of these are proprietary.

### 13.3    CLASSIFICATION OF PLASTICS

Plastics are classified into 3 main groups: thermoplastics (thermoplasts), thermosetting resins (thermosets or resins) and elastomers. The classification is based on their chemical structure and hence properties.

### 13.3.1 Thermoplastics

These are chain molecules in which covalent bonds hold the atoms along the chain together. The chains are however held to each other only by weak secondary bonds. At low temperatures the secondary bonds can overcome the effects of thermal agitation and hence the polymer is frozen into a glassy state. As the temperature increases, the secondary bonds are overcome and segments of the chain can then vibrate relative to each other. The temperature at which this happens is termed the glass transition temperature, $T_g$. If the specific volume of a polymer is plotted against the temperature, a change in slope occurs at the glass transition temperature as shown in figure 13.5. As the temperature is increased further, more of the secondary bonds are overcome until whole chains can slide against each other and the polymer "flows". The temperature at which the material starts flowing is termed the "melting" point, $T_m$.

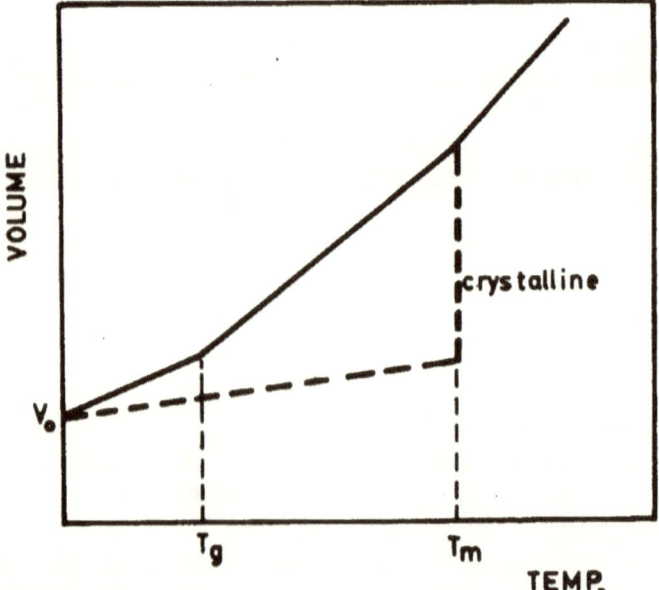

**Fig. 13.5** The variation of specific volume with temperature in thermoplastics

In amorphous polymers, the material behaves like a very viscous liquid between $T_g$ and $T_m$. The material can therefore be thought of a super cooled liquid. The basic structure of the material is not changed between $T_g$ and $T_m$ (the primary bonds are still intact) hence thermoplasts can be formed into desired shapes in this temperature range. On cooling, the secondary bonds are re-established. As a consequence, thermoplastics exhibit thermal-reversible properties.

There may be variations in the temperature versus specific volume curve shown in figure 13.5 due to:

(i) Cooling rate. $T_g$ depends on the cooling rate. It is higher (apparently) when the rate of cooling is higher. For this reason, the glass transition temperature is sometimes termed the kinetic glass transition temperature.

(ii) Crystallinity. An ideal crystalline polymer with one molecular weight does not show a glass transition. Instead, it has a sharp well defined melting point (accompanied by a sharp increase in volume when the crystals change to amorphous liquid). Partially crystalline polymers melt over a range of temperatures. However, the temperature range is narrower compared to fully amorphous ones.

Thermoplasts form the bulk of general engineering plastics and include polyethylene, polypropylene, polystyrene, PTFE, PVC, PMMA and nylon 66. The monomers from which thermoplasts form are bifunctional and the polymers are formed by addition polymerization.

## 13.3.2 Thermosetting Resins

These are also called network polymers. Here all bonds are primary. What results is a three-dimensional network of covalently bonded atoms in which it is impossible to isolate a single chain. Thermosets remain strong and rigid with increase in temperature up to chemical dissociation. They cannot be fabricated by plastic forming. In their manufacture, we start with the partially polymerized molecules and once polymerization is complete, the process cannot be reversed. Thermosets cannot therefore be re-circulated. To form thermosets at least one of the starting monomers must be polyfunctional and in general, the polymer forms by condensation polymerization. Furthermore, thermosets are generally amorphous due to the difficulty of arranging the complex networks into any form of order.

Examples of thermosets are epoxy, polyester, phenol formaldehyde, urea formaldehyde and melamine formaldehyde. Their structures were given in section 13.2.2.

## 13.3.3 Elastomers

These are in between thermosets and thermoplasts. Thus they have a few cross--links (covalent bonds between chains) and secondary bonds between the chains. Normally their glass transition temperature is below room (application) temperature. Their most distinguishing characteristic is their possession of "rubber elasticity" i.e., they remain elastic up to strains in excess of 100 %. The cross links, introduced during their synthesis, provide the "memory" and enable them to come back to their original shape. The common elastomers have the general form:

```
 H         H
 |         |
--C--C==C--C--
 | |   | |
 H H   R H
```

If:

R = CH₃ the result poly(isoprene) or natural rubber

R = H the result is polybutadiene

R = Cl the result is polychloroprene or neoprene, an oil resistant synthetic rubber.

The cross linking can be artificially induced. In this, use is made of the double bond that still exists in the polymer. This process is termed vulcanization and in rubber for example, the cross-linking is provided by sulphur.

Elastomers tend to thermosets as the degree of cross-linking is increased.

## 13.4   MECHANICAL PROPERTIES
### 13.4.1   Introduction

Eventually, the engineer must be able to design with polymers. That is he must specify material and dimensions that will satisfactorily support stated forces. Unfortunately, the mechanical properties of polymers vary very greatly with:

(1) Small changes in temperature around room temperature. Thus, the Young's modulus of a polymer may vary by a factor of $10^3$ or more for a temperature variation from -20 °C to 100 °C.

(2) The time over which the load is applied. Due to creep, the stiffness decreases with time of application of the load.

In addition to the change in magnitude of the properties with temperature and time, there is also a change in the nature of the response. This may change from brittle-elastic, to plastic, to visco-elastic (leathery), to rubbery and finally to viscous flow in a temperature range of 200 °C around room temperature. Further complication comes from the dependence of the properties of a given polymer, even at a constant temperature and time, on the molecular weight, degree of crystallinity, degree of cross linking, etc. The mechanical properties of polymers can therefore not be taken from a handbook and used as easily as those of metals. The variation of two important properties: stiffness and strength, with temperature are considered in more detail below.

## 13.4.2 Time/Temperature Variation of Stiffness

As stated above, Young's modulus in polymers is a function of both temperature and time. Therefore, we can define a relaxation modulus, $E_r$ as the modulus at constant strain, $\varepsilon_o$. i.e.,

$$E_r = \frac{\sigma(t)}{\varepsilon_o} \qquad (13.8)$$

Alternatively, a creep modulus, $E_c$ defined as the modulus at a constant stress, $\sigma_o$, may be specified as:

$$E_c = \frac{\sigma_o}{\varepsilon(t)} \qquad (13.9)$$

Figure 13.6 shows the variation of E (time = constant) with $^T/_{Tg}$, of a typical amorphous thermoplastic. It shows that the response has several stages as stated earlier while the modulus changes from about 3 GPa to virtually zero. The regions are classified as:

(i) Glassy region (elastic-brittle response). This occurs at low temperatures, when both the secondary and primary bonds in the polymer are hard and brittle. It shows a high modulus that varies little with temperature (the little variation is due to chain segments re-adjusting their positions due to thermal energy hence resulting in strain). Any strain is due to bond stretching, which is recovered on removal of load. The bond stiffness can be calculated from composite theory (see chapter 15). If the fraction of primary bonds is f, then:

$$\frac{1}{E} = \frac{f}{E_1} + \frac{1-f}{E_2} \qquad (13.10)$$

where $E_1$ is the modulus of the primary bonds and $E_2$ is the modulus of the secondary bonds. The covalent bonds are similar to the bonds in diamond hence $E_1 = 10^3$ GPa the value of E for diamond. The secondary bonds are similar to those in simple hydrocarbon e.g., paraffin wax, hence $E_2 = 1$ GPa. Thus if $f = 1/2$ (as in most linear polymers), E is approximately 2 GPa.

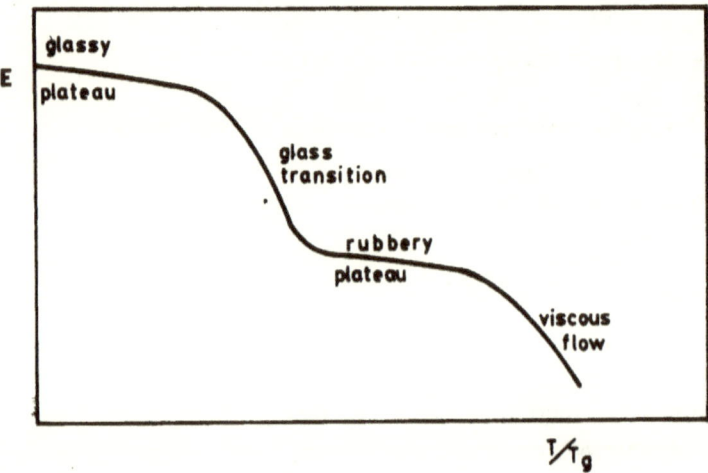

**Fig. 13.6** Variation of Young's modulus with temperature for thermoplasts

(ii) Glass transition region. This occurs when the normalized temperature, $T/T_g$ is about one. Segments of chains start slipping past each other since some secondary bonds have started breaking. Some of the bonds are however, still intact and some strain can be recovered on removal of the load. This recovery is however, resisted by the viscosity of the other chains, and hence recovery continues long after the load is removed. The polymer thus has a leathery behaviour. The strain increases (hence the modulus decreases) with time since chain slipping continues as long as there is load. Thus the polymer exhibits both time independent (elastic) and time-dependent (viscous) characteristics. It is said to be visco-elastic. The visco-elasticity is said to be linear if the strain/stress ratio is a function of the time (or temperature) only.

In the visco-elastic range, we have seen that E changes due to both temperature and time. Either increasing the temperature or increasing the time of application of the load can achieve a given decrease in E. This effect is termed the time/temperature equivalence. A semi-empirical relationship, the William, Landel, and Ferry (WLF) equation, developed using free energy theories for polymers shows that:

$$\log\frac{\tau_o}{\tau} = \frac{C_1(T_1 - T_o)}{C_2 + [T_1 - T_o]} \qquad (13.11)$$

which relates the measurement temperature, $T_1$ with a characteristic relaxation time, t. Usually, the reference temperature, $T_o$ is taken as the glass transition temperature, $T_g$. The WLF equation is valid for amorphous polymers in the temperature range less than 100 K above $T_g$. The constants $C_1$ and $C_2$ have the values $C_1 = -17.44$, $C_2 = 51.66$ K. An alternative equation, the Vogel, Fulcher and Tamman (VFT) equation proposes:

$$-\log \tau(T) = A - \frac{B}{T - T_o} \qquad (13.12)$$

where A and B are other constants. The two sets of constants are related: A $= \log t_g + C_1$, $B = C_1C_2$, and $T_o = T_g - C_2$.

The importance of the time temperature transformation (Equation 13.11) is at once apparent since log (t/$t_g$) gives a time or frequency shift factor, log $a_T$ for a given temperature interval, T - $T_g$. This allows for the construction of a master curve for relaxation times. In other words, tests can be conducted over short times at a higher temperature and the resulting value of E used to estimate the value of E for a much longer time instead of performing a long time test to determine the same property.

(iii) Rubbery region. At even higher temperatures, most secondary bonds are now broken. However, the chains are entangled. The points of entanglement act as cross links and hence the polymer acts as an elastomer. As expected, the rubbery plateau is extended when the molecular weight is increased.

(iv) Viscous phase. At even higher temperatures, (T = 1.4 $T_g$) the chains can now slip past each other and the polymer flows like a viscous liquid. E then approaches zero. It is at this temperature that moulding is done.

(v) Decomposition (or degradation) of the polymer takes place when T = 1.5 $T_g$. The polymer may break into the monomer (like PMMA) or into many different products (like PE).

### 13.4.3 Modulus Diagram for Polymers

The above information can be summarized into a modulus diagram like that shown schematically in figure 13.7. The effects of cross-linking and crystallinity are as shown in figure 13.8.

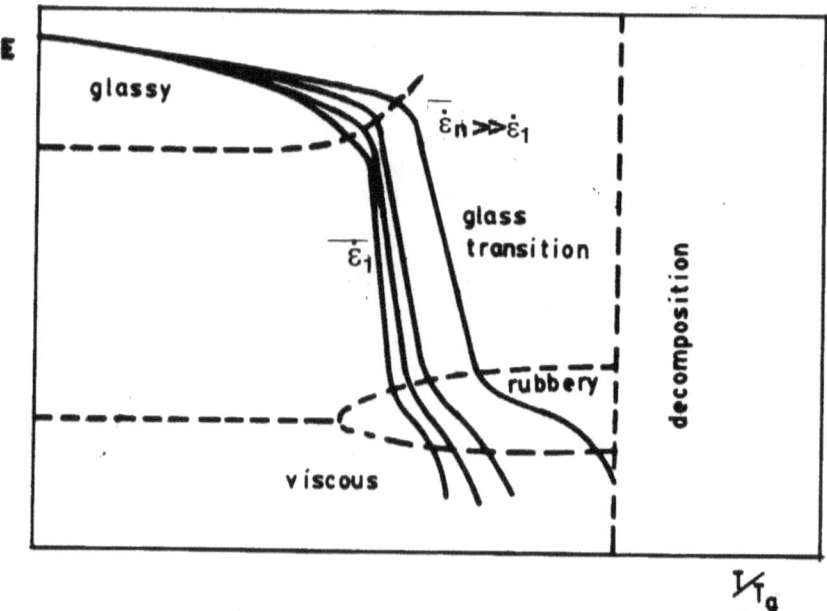

**Fig. 13.7** A schematic showing the variation of modulus with temperature

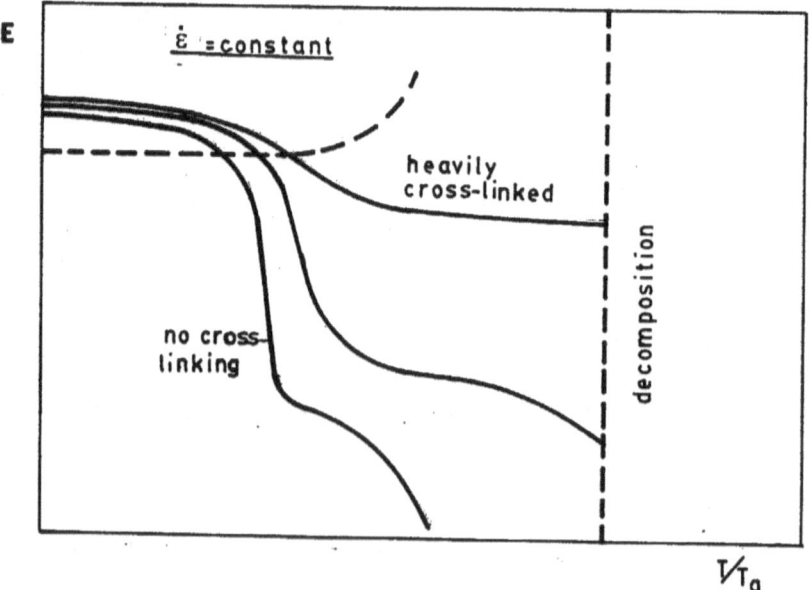

**Fig. 13.8** Effect of cross-linking on the modulus diagram

## 13.4.4    Time/Temperature Variation of Strength

Depending on the normalized temperature, $T/T_g$, a polymer may fail by brittle fracture, cold drawing, shear banding, crazing or viscous flow. Brittle fracture occurs when the normalized temperature is below 0.75. Brittle cracks propagate when the applied stress intensity factor, K, reaches $K_{Ic}$ (or $K_c$) of the material. For polymers, $K_{Ic}$ is approximately 1 MN/m$^{3/2}$.

Cold drawing occurs when T is about 50 K below $T_g$. The stress strain curve then has the form shown in figure 13.9. The polymer yields, starts necking and then draws out as the chains straighten and align themselves to the direction of the stress. The alignment starts at the neck. This makes the neck stronger than the rest of the material and hence the neck spreads rather than continuing to narrow resulting in drawing out the material. The drawing takes place at a fairly constant stress as shown in figure 13.9. When the alignment is complete, a rapid increase in stress is observed and finally failure takes place.

Crazing occurs when the temperature is not low enough for brittle fracture to take place and yet not high enough for plastic deformation (cold drawing) to occur. Small regions are drawn but are constrained by the surrounding. So we have a series of drawn portions (crazes) acting as ligaments joining unaffected material. The crazes appear as white streaks (you can demonstrate this by bending a plastic pen cup). In extreme cases, the crazes act as crack nuclei leading to fracture. The stress strain diagram in this region is as shown in figure 13.10.

Shear banding occurs when the material is loaded in compression at those temperatures at which cold drawing would have taken place in tension. Shear bands occur in the material and the material becomes thicker.

The above information may also be summarized in a strength diagram for a polymer similar to that shown in figure 13.11. But in this case the diagrams are less exact and their use is restricted due to the following factors:

(a) The dependence of the strength on the strain rate.

(b) The complex response of polymers to multi-axial states of stress.

(c) The effects of environment on the properties of polymers. Even innocent looking environments like sunlight may cause embrittlement or degradation in certain polymers thus completely changing their behaviour. The interested reader is referred to the books given at the end of the chapter for more detailed treatment of these aspects.

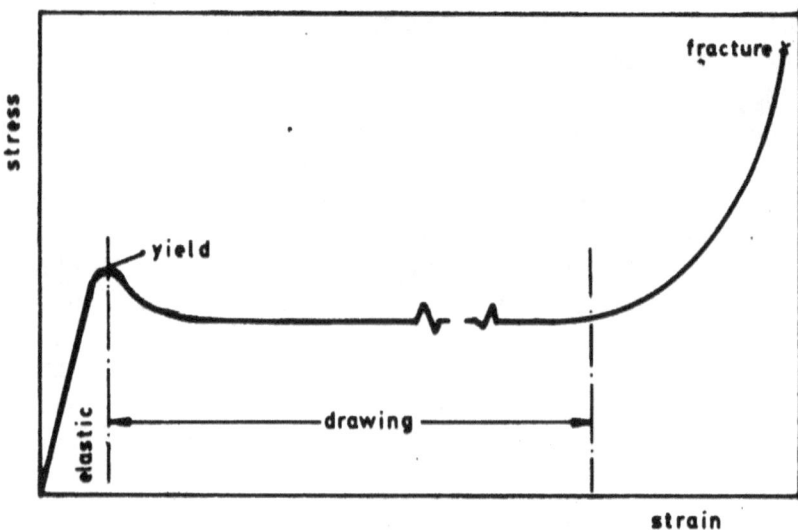

**Fig. 13.9** Stress--strain curve for cold drawing

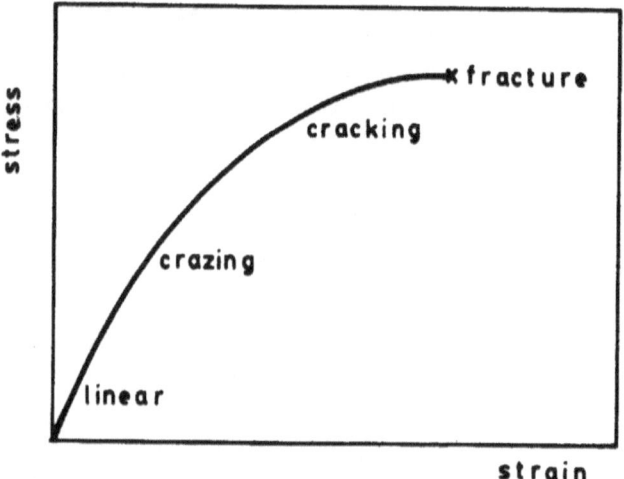

**Fig. 13.10** Stress-strain curve for crazing

## 13.4.5　Mechanical Models

Most polymers are in their visco-elastic regions at room temperature. To better understand the mechanical behaviour in this region, mechanical models are used. The models consist of springs and dashpots. The spring represents the elastic response of the polymer as we have seen i.e., the stretching of the bonds, straightening of zigzag chain between points of cross-linking or entanglement. These displacements are immediately recoverable on removal of the load. The governing equation for the spring is:

$$\sigma = E\varepsilon \tag{13.13}$$

The viscous response of the polymer i.e., the unrecoverable slipping of chains past each other against the viscous resistance of the other chains is represented by the dashpot. The governing equation for the dashpot is:

$$\sigma = \eta^{d\varepsilon}/_{dt} \tag{13.14}$$

where $\eta$ is the viscosity and $^{d\varepsilon}/_{dt}$ is the strain rate.

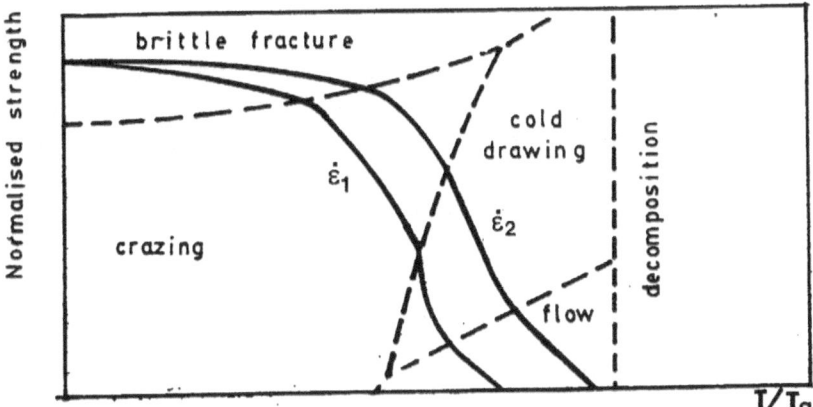

**Fig. 13.11** Schematic representation of a strength diagram for a linear polymer

Several models have been proposed for the analysis of linear visco-elasticity. Some of these are:

## The Maxwell Model

Here, the spring and dashpot are depicted as being in series as shown in figure 13.12 (a). In this case, the stress across each of the elements is the same while the total strain $\varepsilon$ is the sum of the individual strains in each element i.e:

$$\varepsilon = \varepsilon_e + \varepsilon_v \tag{13.15}$$

where $\varepsilon_e$ is the elastic strain or strain in the spring, while $\varepsilon_v$ is the viscous strain i.e., strain on the dashpot.

But:    $\varepsilon_e = {}^{\sigma}/_E$

And:

$$\frac{d\varepsilon}{dt} = \frac{\sigma}{\eta}$$

$$\Rightarrow \varepsilon_v = \int \frac{\sigma}{\eta} dt$$

Hence

$$\varepsilon = \frac{\sigma}{E} + \int \frac{\sigma}{\eta} dt \qquad\qquad (13.16)$$

To model creep (i.e., the variation of strain as a function of time at a constant stress), the substitution $\sigma = \sigma_o$ = constant is made. Here, $\sigma_o$ is the constant stress applied. Then:

On application of the force, all the strain is taken up by the spring, i.e., at $t = 0$, $\varepsilon_v = 0$ and $\varepsilon_e = \varepsilon_o = \sigma_o/E$. Hence:

$$\varepsilon_v = \frac{\sigma_o}{\eta} t + C$$

At $t = 0$, $\varepsilon_v = 0 \Rightarrow C = 0$. Therefore:

$$\varepsilon = \frac{\sigma_o}{E} + \frac{\sigma_o}{\eta} t \qquad\qquad (13.17)$$

The variation of strain with time (creep) of a Maxwell model is therefore as shown in figure 13.12 (b).

To model stress relaxation (i.e., the variation of stress with time at a constant strain), the substitution, $\varepsilon = \varepsilon_o$ = constant is made in equation 13.16. Then:

$$\varepsilon_o = \frac{\sigma}{E} + \int \frac{\sigma}{\eta} dt$$

Since strain is constant, ${}^{d\varepsilon}/_{dt} = 0$. Hence:

$$\frac{d\varepsilon}{dt} + \frac{E}{\eta}\sigma = 0$$

This is a first order differential equation with the general solution:

$$\sigma = Be^{-\frac{E}{\eta}t} \qquad (13.18)$$

where B is a constant of integration. At t = 0, all the strain is taken by the spring since extension of dashpot = 0. Hence, $\sigma_o = \varepsilon_o \times E \Longrightarrow B = E\varepsilon_o$. Therefore:

$$\sigma = E\varepsilon_o e^{-\frac{E}{\eta}t} = E\varepsilon_o e^{-\frac{t}{\tau_R}} \qquad (13.19)$$

The factor $\tau_R = {}^{\eta}/_E$ is termed the relaxation time.

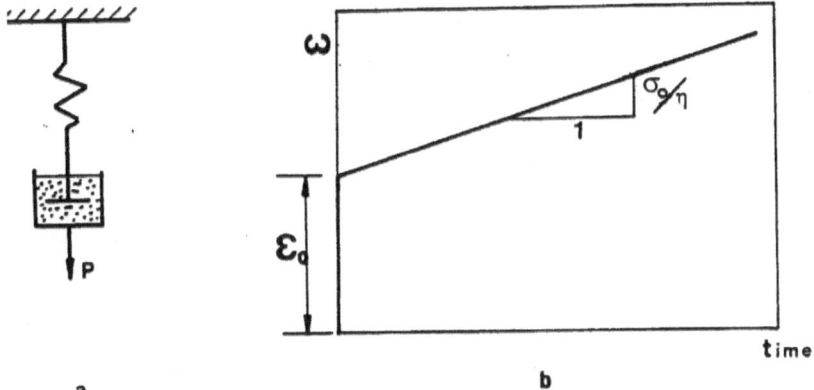

a      b

**Fig. 13.12** The Maxwell model (a) the model (b) the resulting creep curve

The relaxation of a Maxwell model is therefore as shown in figure 13.13.

## The Voigt-Kelvin Model

The spring and dashpot are depicted here as being in parallel as shown in figure 13.14. Hence the strain of both elements is the same while the total stress is the sum of the stresses in each element i.e:

$$\sigma = \sigma_e + \sigma_v$$

$$\therefore \sigma = E\varepsilon + \eta\frac{d\varepsilon}{dt} \qquad (13.20)$$

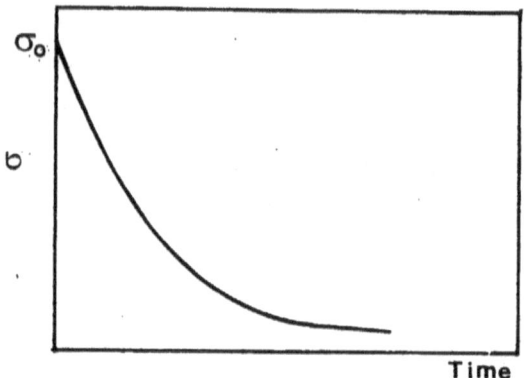

**Fig. 13.13** Stress relaxation of the Maxwell model

To model creep, set $\sigma = \sigma_o = $ constant, Then:

$$E\varepsilon + \eta\frac{d\varepsilon}{dt} = \sigma_o$$

$$\Rightarrow \frac{d\varepsilon}{dt} + \frac{E_o}{\eta}\varepsilon - \frac{\sigma}{\eta} = 0 \qquad (13.21)$$

This is a first order differential equation with the general solution:

$$\varepsilon = \frac{\sigma_o}{E} + Be^{-\frac{E}{\eta}t} \qquad (13.22)$$

where B is a constant of integration. At $t = 0$, $\varepsilon = 0$ since the dashpot cannot extend $==> B = -\sigma_o/E$. Therefore:

$$\varepsilon = \frac{\sigma_o}{E}[1 - e^{-\frac{E}{\eta}t}] \qquad (13.23)$$

The resulting creep curve is shown in figure 13.14 (b). The Voigt-Kelvin model does not allow modeling of stress relaxation.

## The Maxwell-Voigt (or Burgers) Model

This model, also termed the four-element model, is a series combination of the Maxwell and the Voigt-Kelvin models (figure 13.15). The creep equation for the model may be obtained by superimposing equations 13.17 and 13.23. As expected, this model predicts the behaviour of polymers more closely than the first two. Even better approximation may be obtained by having several such models in series and parallel but of course at the cost of increased mathematical complexity.

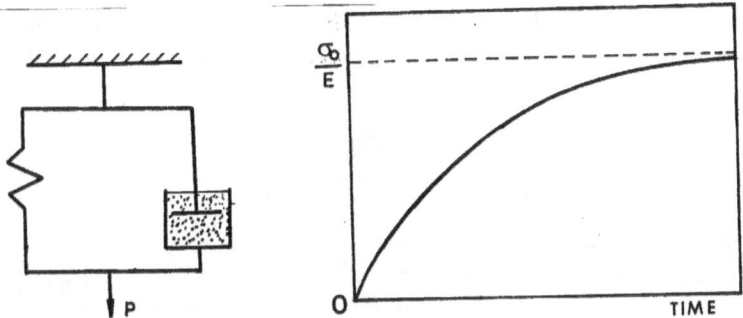

**Fig. 13.14** The Voigt--Kelvin model (a) the model (b) the creep curve

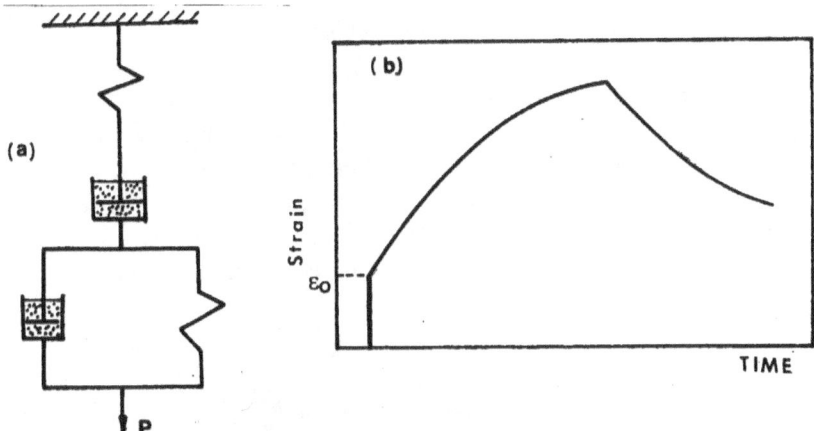

**Fig. 13.15** The Maxwell-Voigt model and the corresponding creep curve

## 13.5   POLYMER PROCESSING

Polymers can be processed and joined by a variety of methods including extrusion, injection moulding, welding, etc. Some of these methods are briefly reviewed hereafter.

### 13.5.1   Extrusion

This method is used mainly for processing thermoplasts though it may also be used for thermosets and elastomers. The plastic is forced through a die opening by a screw and hence extruded into tubes, rods, sheets, etc. In figure 13.16, the raw material (granules of the thermoplast or the two components making up a thermoset) is placed into the hopper from which it falls into the cylinder. The screw pushes the material through the heated part of the cylinder where it is softened before being extruded through the die. As it leaves the die, it is cooled and hardens. The process is rapid and economical, and is the process used to apply insulation to electric cables.

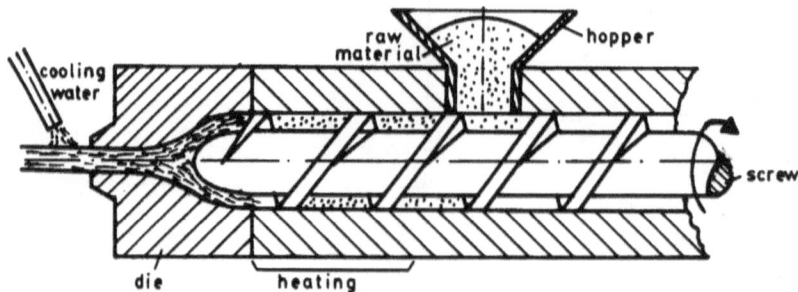

**Fig. 13.16** Schematic representation of extrusion

## 13.5.2    Injection Moulding

This process is depicted in figure 13.17. The plunger is drawn back allowing the raw material (again in the form of granules of the polymer) to fall into the cylinder. The plunger is then pushed forward, forcing the material through the heating cylinder. Further increase of the force pushes the softened material into the mould. The component is allowed to cool while still in the mould.

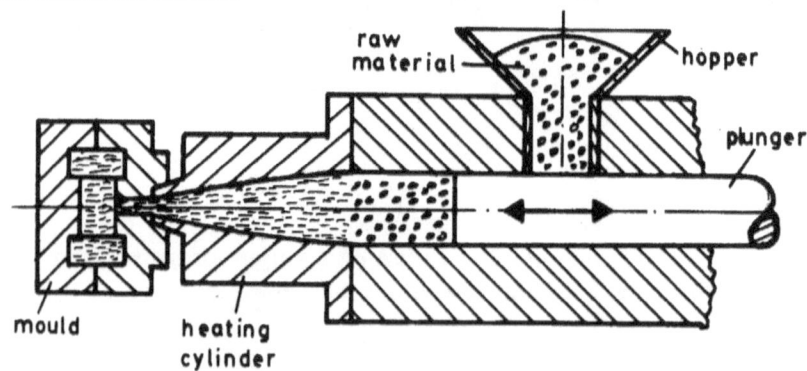

**Fig. 13.17** Schematic representation of injection moulding

Injection moulding has the following advantages:
  (i)    It is easy to automate
  (ii)   Large volume production is possible
  (iii)  Close tolerances can be achieved
However, it suffers from the following disadvantages:
  (a)  The initial capital investment in the machinery and moulds is high.

(b)  Critical control of the process is necessary.

(c)  There is some wastage in the runners.

### 13.5.3    Compression and Transfer Moulding

These two processes are used primarily for thermosets. In compression moulding (figure 13.18), the correct quantity of polymer is placed into a heated half mould. The second half of the mould is then forced onto it. The part is allowed to cool and harden while still in the mould. Usually, the starting materials are the components of a thermoset that set and harden under heat and pressure. The method has the advantage that virtually no waste is produced and the moulds and machinery require little capital investment. But its disadvantages are that the cycle times are long (pressure has to be applied until the part sets) and that thick sections cannot be made.

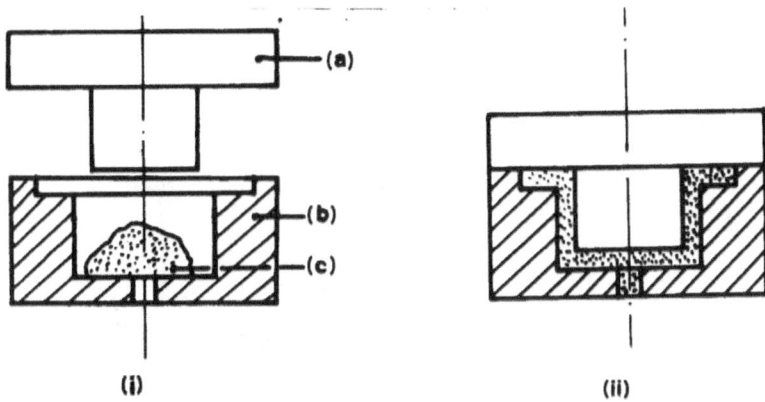

(i)                                                                 (ii)

**Fig. 13.18** Compression moulding (a) before closing mould (ii) after closing mould (a) press (b) heated $1/2$ mould (c) polymer

The procedure is similar in transfer moulding but now the heating takes place in a heating chamber and the softened material is forced into a mould cavity as shown in figure 13.19. Thicker sections can now be made and the cycle times are shorter but the scrap rate and cost of moulds is high.

### 13.5.4    Thermoforming

This method can be used only for thermoplasts. Starting from a sheet, the material is heated and softened and then pressed against the desired shape. In this manner, plastic plates, trays, bowls and other thin walled products are made.

Two common variations to the mechanical forming described above are vacuum forming, where the heated plastic sheet is forced into the desired shape by creating a vacuum under the sheet as shown in figure 13.20. In blow moulding (extensively used for making plastic bottles) a heated length of tube (called a parison) is placed on the air nozzle in a mould. Air is blown into the tube and forces the tube against the walls of the mould as shown in figure 13.21.

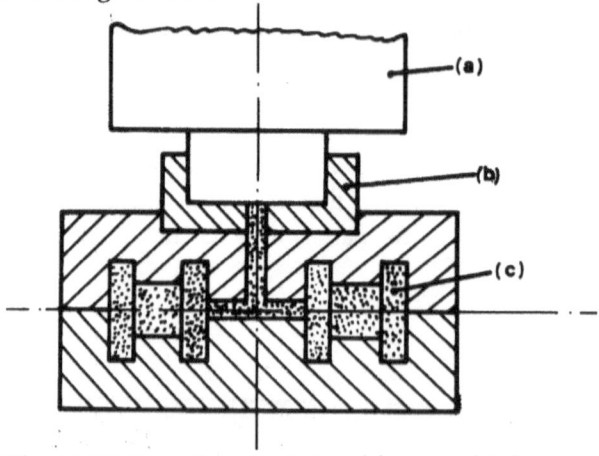

**Fig. 13.19** Transfer moulding (a) press (b) heating chamber (c) mould cavity.

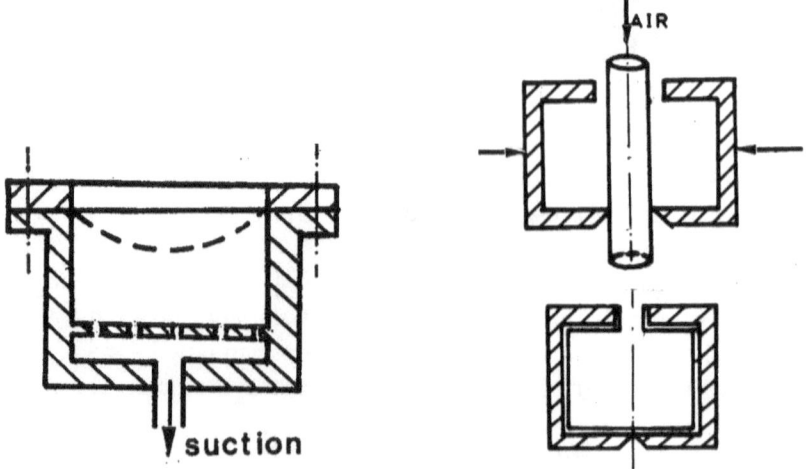

**Fig. 13.20** Vacuum moulding          **Fig. 13.21** Blow moulding

### 13.5.5  Foaming and Filament Winding

Foams are made by expanding a fluid polymer into a low-density cellular structure. An extruder feeds the melt, containing the foaming agent (compressed $CO_2$, or volatile liquid) into an accumulator. The melt is then discharged into a mould (which is at lower a pressure). The foaming agent evaporates leaving a cellular structure. The most common products made by this method are polyurethane foam used in domestic mattresses, cushions, etc., and PS packaging products.

Filament winding is used to make plastic films and fibers. The material is extracted through fine slits. If this process is accompanied by tensioning, aligned fibers (i.e., fibers in which the polymer chains are aligned in the axial direction) are produced. In this manner, high strength polymer fibers like PP, Kevlar (made by Du Pont Company) or Monsanto X-500-G (from Monsanto) are made.

### 13.5.6   Polymer Joining and Strengthening

Polymers may be strengthened by changing their chemical nature. Copolymerization, stabilization (addition of stabilizers) and vulcanization, which we have already considered are methods of chemical strengthening. There is also the possibility of forming two phase "alloys" similar to those in metals. The best example is high impact polystyrene, which is an "alloy" of styrene and butadiene. In manufacture of the "alloy", styrene and butadiene are mixed and heated together. The result is a blend of polystyrene and a rubbery copolymer of styrene and butadiene. The copolymer precipitates inside a polystyrene matrix thus helps in arresting cracks and hence preventing crack growth. Two-phase alloys are formed in polymers because of partial miscibility of the two components. The result is polystyrene that is tougher, an effect similar to the role of precipitates of $Fe_3C$ in improving the toughness of steel.

Joining of polymers can usually be avoided by an integral design. When it (joining) must be done, it can be achieved by cementing, welding or fastening. Cementing is done by using a solution of the same polymer in a volatile solvent. When applied to two surfaces, which are then brought together, the solvent evaporates leaving the two pieces bonded. Alternatively, adhesives (usually epoxy resins) may be used. Welding of polymers is achieved by friction welding. Here, the two surfaces to be welded are moved relative to each other. The friction heat generated softens the materials, which can then fuse if left to solidify under a load. Polymers may also be fastened together. Usually, threads are moulded into one of the parts to allow use of a bolt or screw.

## PROBLEMS

13.1 Sketch and explain the structures of the three basic groups of plastics. How do the structures above account for the mechanical properties of each of the groups at various temperatures?

13.2 (a) What are the major advantages that have contributed to the increased use of plastics as engineering materials?

(b) Explain the meaning of the term "copolymer" and give 4 different types of copolymers.

(c) Sketch and indicate briefly the two types of molecular arrangements in thermoplastics and explain the influence of these on the mechanical properties of the plastics.

13.3 What do the terms addition polymerization, degree of polymerization, cross-linking and copolymerization mean?

13.4 Derive an expression for the stress relaxation of a Maxwell model subjected to a constant strain and draw the expected stress-time graph.

13.5 In analyzing the mechanical behaviour of polymers, models are used consisting of springs and dashpots. What response does each element (i.e., spring and dashpot) represent in the actual polymer?

13.6 With the aid of sketches give a brief a description of the processes of compression moulding and injection moulding. Name two advantages and two disadvantages of each process.

13.7 Explain what is meant by copolymer and homopolymer. What are the different arrangements possible in copolymers?

13.8 Show, on a sketch, the variation of volume with temperature that you would expect when a thermoplastic is heated, and explain the shape of the graph. How would the behaviour of a thermosetting resin differ from the above?

13.9 Derive an equation for the creep behaviour of a polymer based on the Voigt-Kelvin model and sketch the strain-time graph that would be expected.

13.10 (a) Describe with the aid of sketches, how models may represent the mechanical properties of polymers made up of springs and dashpots.

(b) A polymer, whose mechanical properties are closely approximated by the Voigt-Kelvin model has a relaxation time of 70 days, and a Young's modulus of the elastic response, E = 100 MPa at 25°C. It is subjected to a constant stress of 8 MPa at this temperature. Calculate the strain in the material after 60 days. Derive any formula used. (Ans. 0.046).

## Further Reading

Williams, D.J. Polymer Science and Engineering, Prentice-Hall, 1971.

Billmeyer, F.W. Textbook of Polymer Science, 3rd ed. Willey Interscience, 1984.

Brydson, J.A. Plastic Materials, 4th ed. Butterworth, 1982.

Crawford, R. J. Plastics Engineering, 3rd ed. Elsevier, 2004.

# CHAPTER FOURTEEN

# INTRODUCTION TO CERAMICS

## 14.1   INTRODUCTION

Broadly speaking, ceramics are compounds between metals and non-metals (mainly oxides, nitrides or carbides). The bonds are mainly or totally ionic. They are probably the oldest materials used by man and include such common items as pots, china, porcelain, bricks tiles, glasses and all other clay products.

## 14.2   CLASSIFICATION OF CERAMICS

Ceramics may be classified into six groups: glasses, clay products (including white ware), refractories, abrasives, cements, and advanced ceramics. The first three are broadly termed traditional ceramics since they have been used for centuries. The last group has come into widespread use only in the last few decades.

### 14.2.1   Glasses

Glasses are used as containers, in windows, lenses and for manufacture of fibreglass used as reinforcement materials in composite materials. They are generally non-crystalline silicates containing other oxides like CaO (lime), $Na_2O$ (soda), $K_2O$, and $Al_2O_3$. The distinguishing characteristic is their transparency. Addition of other oxides to $SiO_2$ (silica) impacts specific properties to the glass e.g., thermal resistance, chemical resistance, workability, optical properties, etc. Some common glasses are listed below:

|  |  | USES |
|---|---|---|
| Vicor: | 96% $SiO_2$; 4% $B_2O_3$ | Laboratory ware |
| Pyrex: | 81% $SiO_2$; 13% $B_2O_3$; | |
| | 3.5% $Na_2O$;  2.5% $Al_2O_3$ | Oven ware |
| Soda-lime | 74% $SiO_2$; 16% $Na_2O$; | |
| | 5% CaO;  5% others | Glass Containers |
| Fibre glass | 55% $SiO_2$; 16% CaO; | |
| | 15% $Al_2O_3$;  10% $B_2O_3$; | |
| | 4% MgO | In Composites |
| Optical Flint | 54% $SiO_2$; 37% PbO; | |
| | 8% $K_2O$ | Optical Lenses |

## 14.2.2 Clay Products

This group of ceramics uses clay as the raw material. Since clay is abundant in the Earth's crust, these products are relatively cheap. Moreover, the manufacturing process involves only addition of water, adding to the low cost. They may be further divided into two categories: structural products (bricks, tiles, etc.) and white wares (porcelain, pottery, tableware, sanitary ware, etc.).

Clays contain mainly silica ($SiO_2$) and alumina ($Al_2O_3$) plus chemically bonded water. They have a complex crystal structure but all have a layered structure.

## 14.2.3 Refractories

The key properties of refractories are their ability to withstand high temperatures without melting, and their ability to withstand severe environments (i.e., their chemical stability). Moreover, they are able to provide thermal insulation. Their main application (usually in brick form) is for lining of furnaces in metallurgical processes.

Refractories are classified as fireclay ($SiO_2$ plus $Al_2O_3$), silica or acidic ($SiO_2$ plus about 2% CaO), basic (over 90% MgO, $Fe_2O_3$ plus CaO) and the carbides (TiC, SiC, ZrC, and $B_4C$). In addition to these, there is graphite, which is an excellent refractory when used in a non-oxidizing environment but is not a ceramic in the classic definition of a ceramic as a compound between a metal and a non metal. Silica is probably the most often used refractory due to its abundance and low density. Its main disadvantage is the polymorphic change (quartz to tridymite to cristobalite) it undergoes when the temperature changes. These changes may result in thermal stresses. For this reason, slow heating/cooling is required when silica is used in furnaces.

## 14.2.4 Abrasives

Abrasives are used to grind other materials. The main requirement here is high hardness; wear resistance, toughness and high temperature stability. The common abrasive materials are SiC, WC and $Al_2O_3$. Usually, particles of these are bonded to grinding wheels or coated onto paper or cloth.

## 14.2.4 Cements

These include Portland cement, plaster of Paris and lime. When mixed with water, they form a paste, which then hardens over a period of time.

## 14.2.5    Advanced Ceramics

Ceramic materials are finding new uses in electrical, magnetic, optical and engine applications. In particular, the high temperature resistance of ceramics makes them attractive candidates for combustion engines where high operating temperatures translate into high fuel efficiency. Moreover, ceramics have lower density, higher wear resistance as well as lower friction loses when compared to metals. Ceramics used for these purposes are termed advanced ceramics and have potential use even in aircraft engines. Among the advanced ceramics are silicon nitride ($Si_3N_4$), silicon carbide (SiC) and Zirconia ($ZrO_2$). In addition to components made from the ceramics, advanced ceramics are also used as thermal barrier coating in metallic engine parts in aircraft and aerospace applications. Advanced ceramics are also used as armour plate, and in the electronics industry as packaging for integrated circuits. In this application, the low electricity conduction combined with high thermal conduction is exploited. Of particular interest in this regard are boron nitride, aluminium nitride and silicon carbide.

However, before widespread use of ceramics in these applications can take root, toughening mechanisms must be developed to control brittle fracture.

## 14.3    CERAMIC STRUCTURES
## 14.3.1    Introduction

As already stated, ceramics are compounds of metals and non-metals in which the interatomic bonds are predominantly ionic. Ceramics may be either crystalline or amorphous (glassy), but all must have short-range order. As such, each atom must have a specific set of nearest neighbors. Being ionic compounds, ceramic building units are ions rather than atoms. In their coordination, it is necessary that the net electrical charge be zero. Moreover, the sizes of the cations and anions are different. For these reasons, the crystal structures of crystalline ceramics are much more complex compared to those of metals that were considered in chapter two.

## 14.3.2 Ceramics With Long Range Order

The crystal structures of crystalline ceramics are influenced by the magnitude of the electronic charge on each ion (since the crystal must be electrically neutral), and the relative sizes of the cation and the anion. As a rule, anions are larger than cations. Each cation prefers to have as many anions as possible as its nearest neighbours. The most stable structures results when nearest neighbours touch each other. The cation to anion radius ratio determines the coordination number which may be calculated purely from geometric considerations. For very small cations ($r_c/r_A <$ 0.155), the coordination number is two while for $0.155 < r_c/r_A < 0.225$, the coordination number is three (see example 14.1). Note: $r_A$ and $r_c$ are the anion and cation radii, respectively.

Example 14.1: Calculate the minimum cation to anion ratio for a coordination number of three.

Answer:

Consider the case where all anions are in contact with the cation and also with each other:

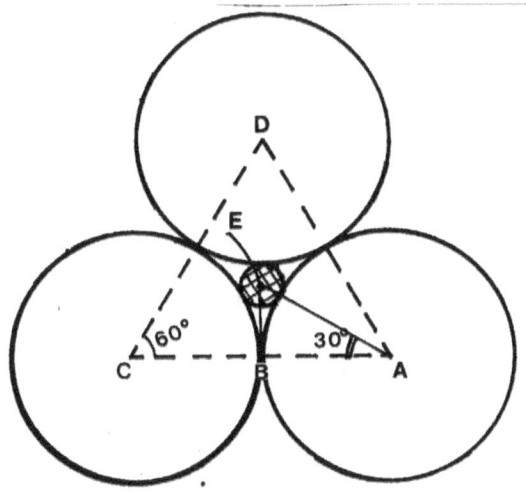

**Fig. E 14.1**

$AB = r_A$;  $AE = r_A + r_c$;    $/\_ DAC = 60°$; $/\_ BAE = 30°$

$AB = AE \cos 30° =>$    $r_A = (r_A + r_c) \times 0.866$

$=> r_c = 0.155 r_A$

Similarly, it can be shown that $r_c/r_A$ ratios between 0.225 and 0.414 favour a coordination number of four; 0.414 to 0.732 a coordination number of six; 0.732 to 1.0 a coordination number of eight while a ratio larger than one favours a coordination number of twelve. If the coordination number is four, the cation resides at the centre of a tetragon formed by four anions. When the coordination number is six, the cation is at the centre of an octrahedron of anions. A coordination number of eight resembles a BCC arrangement in metal crystals with the cation at the cube centre. A coordination number of twelve resembles the FCC arrangement in metal crystals with cations at the face centres.

## (a) A-X Type Crystal Structures

These are crystals in which the number of anions equals that of cations. Within this group, several individual crystal structures are possible. These are usually named after a common ceramic with the said structure as detailed below:

### 1. The Rock Salt Structure

This is the structure of sodium chloride (rock salt), which is shown in figure 14.1.

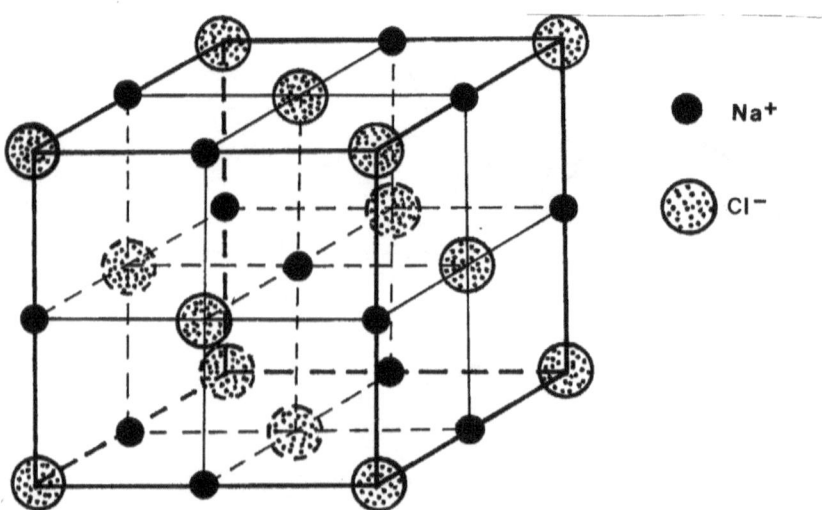

**Fig. 14.1** The rock salt crystal structure

The coordination number is six. There is an FCC arrangement of chlorine ions with cations (sodium ions) at the cube centre and on each of the cube edges. Other ceramics with this structure are MgO, MnS and LiS.

### 2. Cesium Chloride Structure

This is similar to a BCC metal crystal structure but the atom at the centre is replaced with a cation while the anions occupy the cube corners. The coordination number is eight (see figure 14.2).

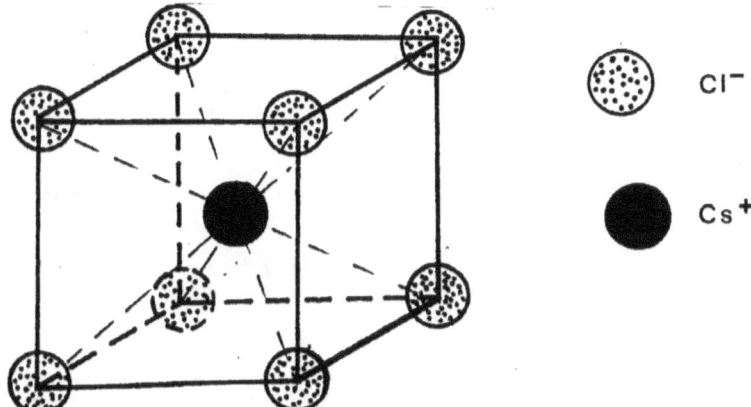

**Fig. 14.2** A unit cell of CsCl

## 3. Zinc Blende Structure

This is the structure of zinc sulphide and is shown in figure 14.3. The coordination number is four with each zinc ion surrounded by a tetragonal arrangement of sulphur ions (anions). Silicon carbide (SiC) has a similar structure.

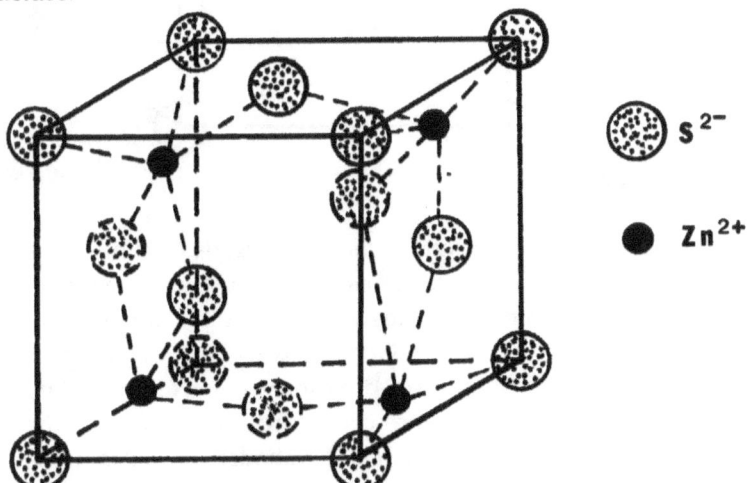

**Fig. 14.3.** The crystal structure of ZnS (zinc blende)

### (b) $A_m X_n$ Type Crystals

In this case, the electrical charge in the cation differs from that on the anion, e.g., $CaF_2$ or $Al_2O_3$. In $CaF_2$ for example, the $r_A:r_c$ ratio favours a coordination number of eight. The structure is similar to that of CsCl with

F⁻ ions occupying the corners of the cube. However, the number of $Ca^{2+}$ is only half that of F⁻ ions meaning that only half of the cation sites (the cube centres) are occupied.

## (c) $A_mB_nX_p$ Type Crystals

There are two types of cations in this case. An example is $BaTiO_3$, which has a cubic structure at temperatures above 120 °C with $Ba^{2+}$ ions taking the eight corners of the cube. A single $Ti^{4+}$ ion occupies the cube centre while $O^{2-}$ ions take the cube faces.

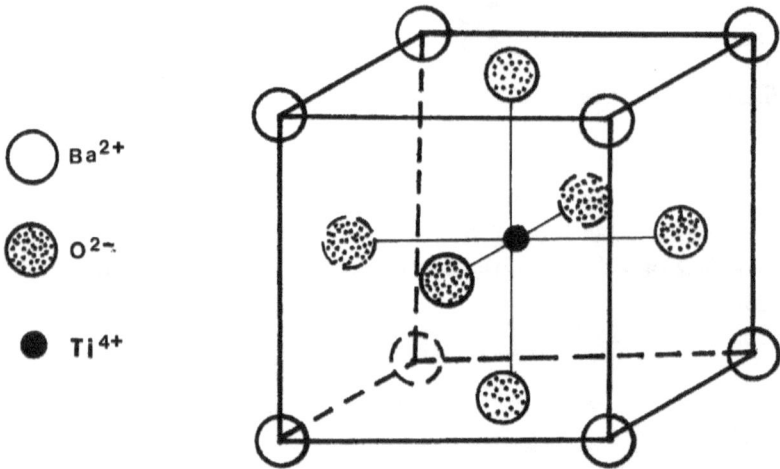

**Fig. 14.4** A unit cell of $BaTiO_3$

Just as in the case of metal crystals, ceramic crystals may be considered as stackings of close packed planes. In this case, the close packed planes are made up of anions. The interstitial sites are taken up by cations. There are two types of interstitial sites: tetragonal positions and octrahedral positions. A tetragonal (or four fold) site if formed by three anions in one plane and one anion from the plane above. Lines drawn from the centres of the four anions form a tetragon hence the name. Octrahedral or six fold sites are formed between three anions in one plane and three others on the plane above. There are two tetragonal sites and one octrahedral site for every anion.

The stacking sequence of the spheres may be the HCP type (ABABAB...) or the FCC type (ABCABCABC...). Ceramic crystal structures depend on this stacking sequence and on the manner in which the interstitial positions are filled with cations. The rock salt structure for example involves an FCC packing of Cl⁻ ions with Na⁺ ions occupying all octrahedral positions.

Example 14.2: Calculate the atomic packing factor of CsCl given that the respective ionic radii are 0.17 nm for Ce⁺ and 0.181nm for Cl⁻.

Answer:

Refer to figure 14.2: $3a^2 = [2(r_{cs}^+ + r_{cl}^-)]^2 = [2(0.17 + 0.181)]^2$

$$\Rightarrow a = 0.41nm$$

There is one Cs⁺ and one Cl⁻ per unit cell of CsCl. Therefore,

$$APF = \frac{vol.\_of\_atoms}{vol.\_of\_unit\_cell} = \frac{4}{3}\pi\frac{[0.17^3 + 0.181^3]}{0.41^3}$$

$$= 0.659$$

Example 14.3: Given that the radius of the ferrous ion is 0.077 nm and that of the oxygen ion is 0.14 nm, predict the crystal structure of the FeO crystal.

Answer:

FeO is an AX type compound:

$$\frac{r_c}{r_A} = \frac{0.077}{0.14} = 0.55$$

The likely coordination number is six
The likely crystal structure is the rock salt structure

Example 14.4: The radii of Ba²⁺ and O²⁻ ions are 0.143 nm and 0.132 nm, respectively. Determine the radius of the interstice into which Ti⁴⁺ ion is placed in BaTiO₃.

Answer:

Refer to figure 14.4, and consider the base of the cube:

$$2a^2 = [2(r_{Ba} + r_O)]^2 = 4(0.143 + 0.132)^2 (nm)^2$$

a = 0.389 nm

If the radius of the interstice is r, then considering a line through the centre of the cube,

$2(r + r_{O2-}) = a = 0.389$ nm

From which $r = 0.0625$ nm

NOTE: Since the majority of crystalline ceramics are minerals, one finds "mineral" names for ceramics in the literature. Some of these names are: rock salt (NaCl), zinc blende or sphalerite (ZnS), Fluorite ($CaF_2$), perovskite ($CaTiO_3$), spinel ($MgAl_2O_4$), silica ($SiO_2$), etc.

## 14.3.3    Silica and the Silicates

All ceramic materials have short-range order i.e., there is a spatial coordination among the atoms. In effect, each atom or ion has a specific arrangement of nearest neighbours. In the crystalline materials considered above, this order is extended to the long range resulting in formation of specific crystals. In some ceramics however, the bulk of the material is amorphous or glassy.

Silica and the silicates form the most abundant group of materials on the Earth's crust. The building unit in these materials is the ion $SiO_4^{4-}$. This group is the basis of all rocks, soils, clays and sand. $SiO_4^{4-}$ exists as a tetragonal arrangement of $O^{2-}$ ions surrounding a $Si^{4+}$ cation. The way the $SiO_4^{4-}$ ions are arranged determines the structure.

When each oxygen ion is, in turn, shared by adjacent tetrahedra, a sort of polymerization takes place and the result is a three dimensional array of tetrahedra with an overall formula $SiO_2$ or silica. The resulting material is electrically neutral and the tetrahedra may be arranged in an ordered manner. This results in a crystalline material, which may exist in three polymorphic forms: quartz, cristobalite or tridymite. The polymorphism is temperature based: quartz is stable below 875 °C, tridymite is stable between 875 °C and 1470 °C, while crystobalite is stable above 1470 °C.

It is also possible for silica to occur without long-range order. This results in vitreous or fused silica. Short-range order still exists in the $SiO_4^{4-}$ unit. Commonly used glasses belong to this class of materials (usually with other oxides e.g., CaO added).

When at least one oxygen ion of the $SiO_4^{4-}$ unit is not shared by other tetrahedra, a net negative charge results. These are neutralized by cations resulting in silicates. Depending on the number of oxygen ions coordinated with other tetrahedra, the resulting silicates will have varying formulae e.g., $SiO_4^{4-}$, $Si_2O_7^{6-}$, $Si_3O_9^{6-}$, etc. The cations perform the additional role of bonding the tetrahedra together (in addition to balancing the electrical charges). An example is $Mg_2SiO_4$ (forsterite). Silicates may also have more than one cation e.g., akermanite ($Ca_2MgSi_2O_7$). Furthermore, sheet structures may sometimes result from polymerization of tetrahedra units in two dimensions, as is the case in clays. In the later case, the repeating unit is $Si_2O_5^{2-}$ where the oxygen ions are shared by other tetrahedra to form a sheet. Unbounded oxygen atoms stick out of such sheets and combine with cations in adjacent sheets. This results in a two-layered plate structure. A simple example of such a clay is kaolinite ($Al_2[Si_2O_5]OH_4$) in which $SiO_5^{2-}$ in one layer is bonded to $Al_2(OH)_4^{2+}$ in another layer. Secondary bonds bond the two-layered unit to similar units.

## 14.4  IMPERFECTIONS IN CERAMIC CRYSTALS

As was pointed out for metal crystals, point defects (vacancies, interstitials, substitutional impurities and interstitial impurities) may exist in ceramic crystals. In this case however, defects may involve either of the ions, i.e., cation or anion (see figure 14.5). However, anion interstitials are rather rare due to the large size of most anions.

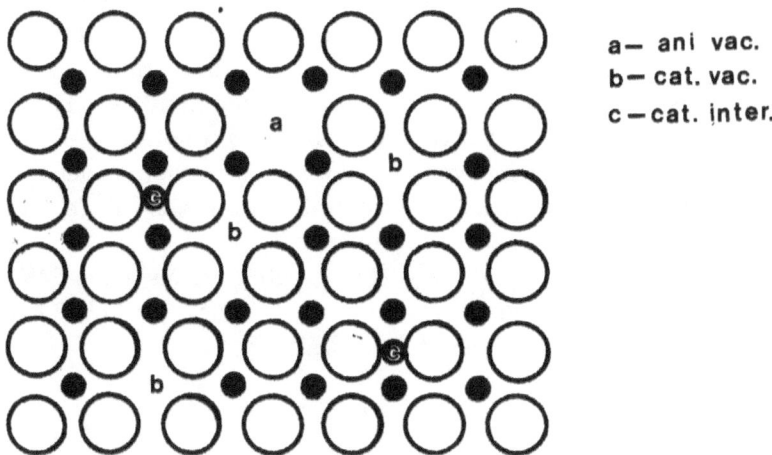

a— ani vac.
b— cat. vac.
c — cat. inter.

**Fig. 14.5** Point defects in ceramics

A further constrain is placed on point defect formation in ceramics by the need to satisfy the charge balance requirement. As a result, defects usually occur as anion-cation pairs. For example, a Frenkel defect involves a cation vacancy and a cation interstitial pair (figure 14.6). A cation may leave its normal lattice site and occupy an interstitial site.

Note that the charge balance is retained. Another "pair" defect is the Schottky defect that involves a cation/anion vacancy pair (figure 14.6). Once again, the charge balance is maintained.

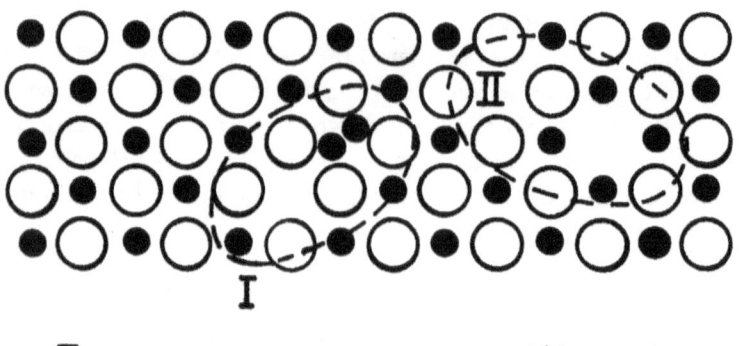

I — **FRENKEL DEFECT**          II — **SCHOTTKY DEFECT**

**Fig. 14.6** Schematic representation of Frenkel and Schottky defects

Point defects in ceramics might also occur due to existence of two valence states (isotopes) in an element. In such cases, an equilibrium ratio between the two isotopes must be maintained. A good example is FeO (wustite). The majority of cations will exist as $Fe^{2+}$, but a finite number of $Fe^{3+}$ ions must also be present to maintain the equilibrium $Fe^{2+} : Fe^{3+}$ ratio. However, the $Fe^{3+}$ disrupts the charge balance. To restore this balance, a cation vacancy must occur for every two $Fe^{3+}$ ions present. But even though the charges are now balanced, nonstoichiometry results, i.e., there are now more anions than suggested by the formula FeO.

Example 14.5: A sample of wustite (FeO) contains w/o 24% $O_2$. Determine the number of vacancies, $Fe^{2+}$ and $Fe^{3+}$ for every 1000 $O^{2-}$ ions. Atomic wt. of oxygen = 16.0; that of Fe = 55.84).

Answer:
Consider 1 kg of FeO
Weight of $O^{2-}$     = 240 g
Weight of $Fe^{2+}$ and $Fe^{3+}$ ions   = 760 g

240/16 is equivalent to:    1000 $O^{2-}$ ions
760/55.84 is equivalent to:    1000 x (760/55.85 x 16/240) Fe ions.
      = 907 Fe ions

If stoichiometry existed, there would be 1000 Fe ions

=> # of vacancies = 1000 - 907

= 93 vacancies

There are two $Fe^{3+}$ ions for every vacancy

=> # of $Fe^{3+}$ ions = 186

=> # of $Fe^{2+}$ ions = 907 - 186

= 721 $Fe^{2+}$ ions

=> $Fe^{3+}$ : $Fe^{2+}$ ratio = 186/721 = 0.258

Both interstitial and substitutional impurity atoms (solid solutions) may form in ceramics. A substitutional impurity will substitute for an ion close to it in terms of electron configuration and size. Thus, a divalent cation will substitute another divalent cation. For example in spinel ($MgAl_2O_4$), $Fe^{2+}$, $Ni^{2+}$ $Zn^{2+}$ or $Mn^{2+}$may easily replace the $Mg^{2+}$. On the other hand, $Fe^{3+}$, $Cr^{3+}$ or $Mn^{3+}$ easily substitutes the $Al^{3+}$.

However, it is possible for ion pairs of different valence to replace another ion pair provided the electrical balance is maintained. As an example, the $Ca^{2+}/Al^{3+}$ pair in anorthite ($CaAl_2Si_2O_8$) may be replaced by a $Na^+Si^{4+}$ pair. Alternatively, a $R^+$ cation may replace an $R^{2+}$ cation if a simultaneous replacement of $X^-$ anion by an $X^{2-}$ anion takes place, or a cation vacancy is created (see the FeO of example 14.5).

## 14.5 CERAMIC PHASE DIAGRAMS

Phase diagrams in ceramic systems are formed both between compounds. Frequently, the two compounds share a common element like oxygen. In such cases, pseudo binary phase diagrams may be drawn similar to those already considered for metallic alloy systems. On occasions however, three or more compounds may be involved in which case ternary and higher order phase diagrams have to be constructed.

### 14.5.1 Pseudo-Binary Phase Diagrams

These are similar to those in metallic alloy systems in all respects except that compounds rather than pure elements are involved. It is therefore possible to have systems with complete miscibility in the solid state, partial solubility in the solid state, etc. As an example, the $Al_2O_3$-$Cr_2O_3$ system shows complete solubility in the solid state and has the same form as the isomorphous nickel-copper system that was considered in chapter five. The reason for this is the similarity between the $Cr^{3+}$ and $Al^{3+}$ ions, which can therefore substitute completely for each other in the crystal lattice.

Moreover, both compounds have the same crystal structure.

The $MgO$-$Al_2O_3$ system on the other hand shows a case of partial solubility in the solid state in which an intermediate compound or phase is also formed. This is the equivalent of an intermetallic compound in metallic alloy systems. Unlike an intermetallic compound, the intermediate phase is non stoichiometric and hence is stable over a range of compositions. In this specific case, the intermediate phase is spinel ($MgAl_2O_4$). The phase diagram is shown in figure 14.7.

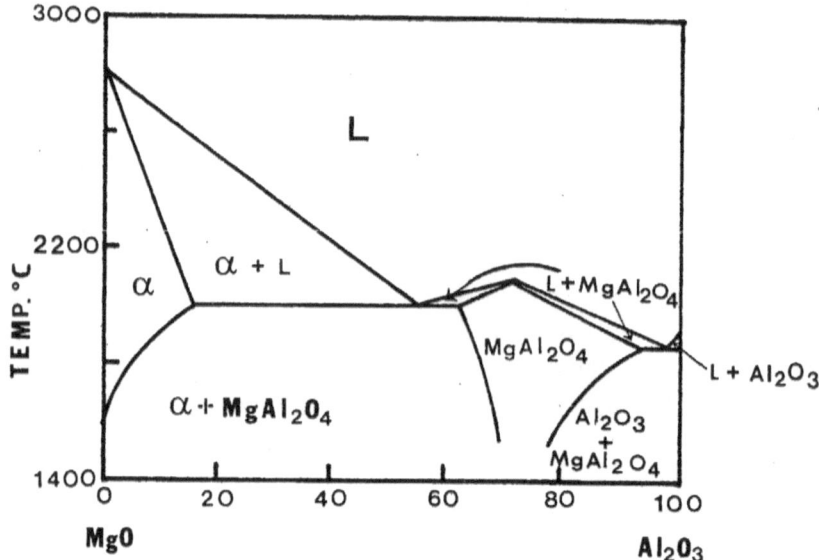

**Fig. 14.7** A schematic representation of the $MgO$-$Al_2O_3$ phase diagram

Peritectic, eutectic and eutectoid reactions do occur in pseudo binary systems as well. A good example can be seen in the $CaO$-$ZrO_2$ (Calcia-Zirconia) system (see figure 14.8). The eutectoid reaction occurs when zirconia changes crystal structure from tetragonal to monolithic at about 900 °C while the peritectic reaction follows the formation of cubic zirconia from tetragonal zirconia at around 2600 °C.

All the factors applicable to binary alloy phase diagrams in metallic alloy systems apply equally well to pseudo binary ceramic phase diagrams. Thus the chemical composition of the phases, ratios of the phases, etc. at any temperature may be determined in the same manner as explained in chapter five.

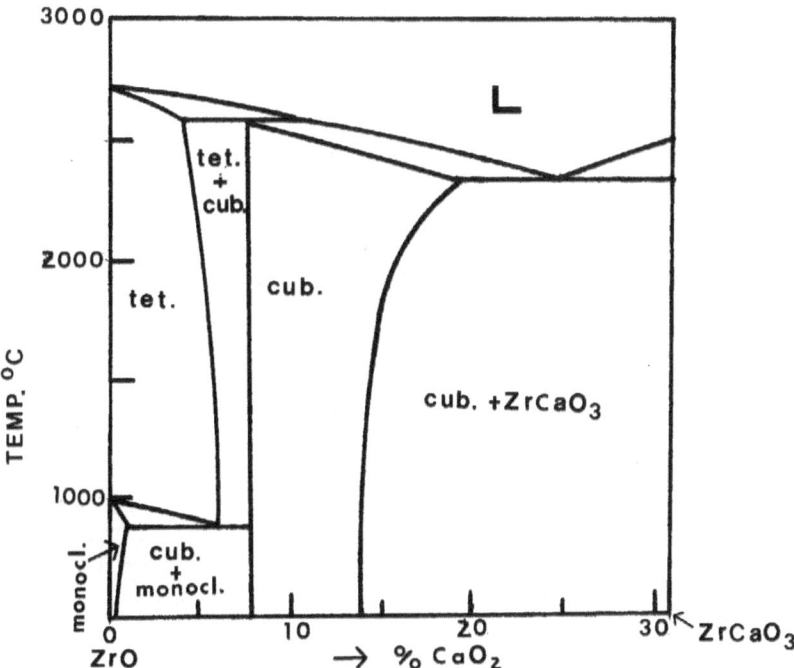

**Fig. 14.8** A schematic representation of the CaO-ZrO₂ phase diagram

### 14.5.3 Ternary Phase Diagrams

Ternary phase diagrams have to be used whenever a system has three distinct components. Due to limitations on the dimensions available for presentation, one more variable has to be fixed. When the temperature for example is fixed, a ternary phase diagram may be drawn as a triangular plot. Each component is represented at one corner of the diagram as shown in figure 14.9.

The lever rule is equally applicable to ternary phase diagrams. For example in figure 14.9, composition X, the ratios are given by:

Ratio of A $= aX/aA$
Ratio of B $= bX/bB$
Ratio of C $= cX/cC$

Moreover: $aX/aA + bX/bB + cX/cC = 1$

A, B and C may be elements, phases in equilibrium or compounds.

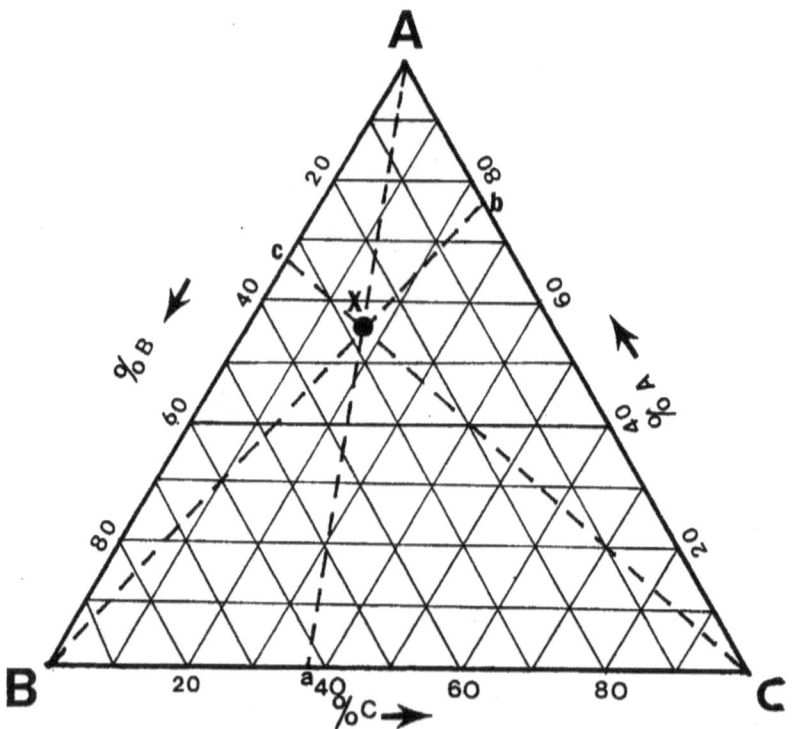

**Fig. 14.9** An isothermal ternary phase diagram with components A, B and C.

An example of an isothermal ternary phase diagram between $SiO_2$ (silica), CaO (calcia) and $Al_2O_3$ (alumina) is shown in figure 14.10. Scope and space limitations do not allow for a more rigorous treatment of ternary phase diagrams and the reader is referred to references at the end of the chapter for more details.

## 14.6 MECHANICAL PROPERTIES OF CERAMICS
### 14.6.1 Strength and Toughness
Generally, ceramics tend to be brittle and to fracture with negligible plastic deformation. In non-crystalline ceramics, the fracture process proceeds by initiation and propagation of cracks in a direction perpendicular to the applied load. In crystalline ceramics, crack growth is transgranular and occurs by cleavage along specific crystallographic planes.

Due to flaws in ceramics (which include microcracks, pores, grain corners, etc.) that act as stress concentrators, the fracture toughness of ceramics is low (about 10 MPam$^{1/2}$ for most ceramics). Ceramics may also undergo static fatigue or delayed fracture. This involves slow crack propagation under a static load. This phenomenon is very environment sensitive. The presence of moisture in the vicinity of the crack is most deleterious leading to the conclusion that the cracking mechanism is stress corrosion cracking. The most affected ceramic is silicate glass.

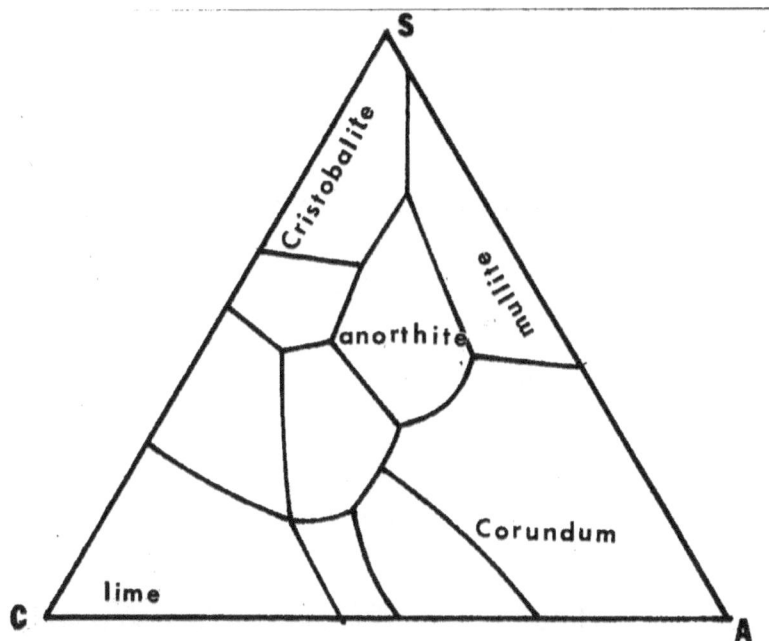

**Fig. 14.10** A ternary phase diagram between silica (S), calcia (C) and alumina showing major minerals derived from the system. Phase boundaries are only approximate while some phases have been omitted for simplicity.

The strength of ceramics is strongly influenced by factors like surface finish and size of the component. As a consequence, there is considerable scatter in strength data collected from a large number of specimens. The reason has already been mentioned above: since the strength depends on flaws, the condition of the surface will dictate the strength. Moreover, the size of the flaws will vary from specimen to specimen accounting for the scatter. For the same reason, large specimens tend to have lower strength than smaller ones due to the higher probability of having large flaws in the latter.

Concise Notes on

The grain size also influences the fracture strength of ceramics in that components with coarser grains are weaker than those with finer grains. This results from the higher stress concentrating effect of coarser grains.

Ceramics are much stronger in compression than in tension since stress raisers do not have any effect in compression. For this reason, ceramics find more frequent use in applications where the loads are compressive. Furthermore, the strength of these materials in tension is enhanced if compressive residual stresses are induced on their surfaces.

## 14.6.2    Stiffness (Elastic Behavior)

Ceramics generally have high stiffness (modulus of elasticity) due to the high interatomic bonds introduced by the ionic bonding. Most values lie between 70 GPa (glass) and 470 GPa (silicon carbide). In single crystals, there is anisotropy in the value of elastic modulus with values in the <100> being higher than those in <110>. The reason for this can be traced to the direction of forces of Coulombic attraction. A strain in the <100> direction tends to separate negative and positive ions and hence meet greater resistance.

Elastic properties in ceramics are determined by transverse bending tests rather than tensile tests due to two reasons:

1. The difficulty that would be involved in producing and gripping tensile specimens with the required profile.
2. The differences in tensile and compressive strengths that has already been alluded to above.

In transverse bending, a bending moment is applied to the specimen through either a four point or a three point bending arrangement. The arrangement for a four point bending test is shown in figure 14.11.

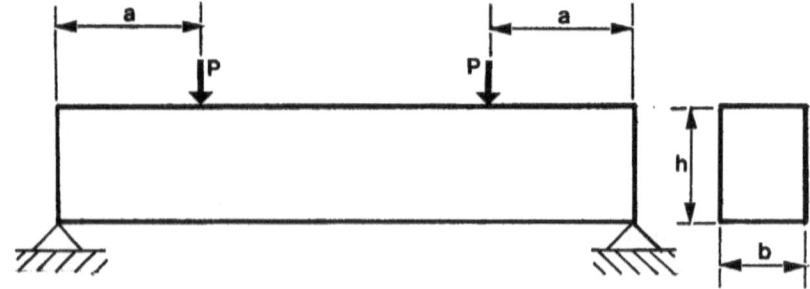

**Fig. 14.11** Four--point bend test arrangement

In such loading, both tensile and compressive stresses are applied through any cross section (a reader, not familiar with such calculations may

250

wish to refer to any standard text on mechanics of materials). The maximum strength at any point between the supports in figure 14.11 is given by:

$$\sigma = Mc/_I \tag{14.1}$$

where,     $M = Pa$
            $c = h/_2$
            $I = bh^3/_{12}$

So that:

$$\sigma = \frac{6Pa}{bh^2} \tag{14.2}$$

The stress at fracture obtained from such tests is termed the modulus of rupture or the bend strength. It is normally higher than the tensile fracture strength but lower than the compressive strength. As explained earlier, the modulus of rupture depends on the specimen size.

### 14.6.3    Hardness

Ceramics are among the hardest materials known. This property makes them suitable for cutting and abrasive tools. Due to their brittleness however, it is necessary to mix different ceramics in grinding wheels before use. As an example, $Fe_2O_3$ and $TiO_2$ are often mixed with $Al_2O_3$ in grinding wheels to improve the latter's toughness.

### 14.6.4    Plastic Deformation in Ceramics

Though plastic deformation in ceramics is limited, some deformation may take place in both crystalline and amorphous ceramics. In layered crystalline ceramics for example, slip may take place under compressive loads. The usual slip system is {110}<110> i.e., slip of {110} planes in the <110> direction. Other slip systems do not operate easily since it would involve bringing like charges together. The scarcity of slip systems explains the brittle nature of ceramics since fracture is likely to precede activation of a slip system.

In non-crystalline ceramics, no slip is possible since there are no slip planes. Plastic deformation takes place by viscous flow. As with viscous flow in plastics, the governing equation is:

$$\eta = \tau / (^{dv}/_{dy}) \tag{14.3}$$

where, $^{dv}/_{dy}$ is the velocity gradient perpendicular to the direction of the stress. The viscosity, $\eta$, decreases with increase in temperature.

### 14.6.5    Effect of Porosity

Due to the methods by which ceramic components are manufactured, they invariably contain some residual porosity. The porosity reduces the stiffness of the ceramic roughly according the formula:

$$E = E_o (1-k_1 p + k_2 p^2) \tag{14.4}$$

where, $E_o$ is the value of E when the porosity is zero, p is the volume fraction of porosity while $k_1$ and $k_2$ are constants. An exponential effect is noted on the strength due to the fact that porosity reduces the cross section area of the specimen in addition to introducing a stress concentration. As such, the modulus of rupture may be estimated by:

$$\sigma_{mr} = \sigma_o exp(-np) \tag{14.5}$$

where $\sigma_{mr}$ is the modulus of rupture, while $\sigma_o$ and n are empirical constants.

## 14.7 THERMAL AND ELECTRICAL PROPERTIES OF CERAMICS

### 14.7.1    Thermal properties

Since there are no free electrons in ceramics, they are poor conductors of heat when compared to metals. Limited heat conduction occurs via phonons i.e., quantinized lattice vibrations. The factors that control conduction are the energy of the phonons, their velocity (which is equal to the velocity of sound in the material), and their mean free path. As a result, crystalline ceramics have a higher heat conductivity compared to glasses. The regularity in crystals ensures reduced interference with motion of phonos. Likewise, some compound crystals have higher heat conductivity than solid solutions.

Heat conduction in ceramics is isotropic only in cubic crystals and glassy ceramics. The anisotropy is most dramatic in graphite where conduction parallel to the hexagonal planes is two orders of magnitude higher than that perpendicular to the planes.

In general, the thermal conductivity decreases with increase in temperature due to higher thermal energy of atoms/ions at the higher temperature. The lattice becomes less regular and interferes more with phonon movement. However, in transparent ceramics, there is an apparent increase of heat conductivity with increase in temperature resulting from radiative heat transfer. The level of porosity and the microstructure also affects the conductivity.

One major problem with ceramics is cracking due to thermal stresses. These stresses are induced by differences in thermal expansion. Ceramics, being brittle, cannot deform plastically to accommodate the differences in strain. This factor leads to thermal cracks. The strain differences may result from one of several reasons:

1. Phase transformation which are usually accompanied by a volume change
2. Differences in coefficient of thermal expansion in multiple phase systems
3. Anisotropy in thermal properties
4. Temperature gradients

For the same reason, stress relaxation occurs in some ceramics when they are held at constant strain.

## 14.7.2 Electrical Properties

Ceramics are poor conductors of electricity due to the lack of free electrons to act as charge carriers. When limited conduction takes place in some ceramics, the mechanism responsible is ionic conduction. Even then, the conductivity is low since the mobility of the heavy ions is low. However, considerable ionic conduction may take place in non-stoichiometric compounds through cation or anion holes. For example in FeO, cation vacancies left by the presence of $Fe^{3+}$ ions allow considerably higher mobility of $Fe^{2+}$ ions.

Under certain circumstances, ceramics may act as semiconductors with the unusual property that conduction increases with increase in temperature. This increase results from greater ion diffussivity at the higher temperature, which more than compensates for the shorter free path lengths. Furthermore, electrons can attain the energy required to move into the conduction band. Semiconducting ceramics are used in fixed resistors, as heating elements (silicon carbide and graphite), as thermisters (temperature sensors) and as rectifiers. Details of these are of more interest to electrical engineers and hence are not considered further.

## 14.8 CERAMIC PROCESSING

Due to the high melting points and brittleness of most ceramics, casting and processes involving plastic deformation are not used for processing ceramics. The major processing methods for ceramics are:

(a) Glass forming
(b) Particulate forming, and
(c) Cementation.

### 14.8.1 Glass Forming

The variation of specific volume with temperature in glasses resembles that of thermoplastics considered in chapter thirteen. In other words, there is no clear solidification point. Instead, the viscosity of the liquid increases gradually as the temperature is decreased. At some point, a decrease in the slope of the curve is noted. The corresponding temperature is the glass transition temperature, $T_g$. Below $T_g$, the material is a super cooled liquid, while above the temperature marked as $T_m$ in figure 14.12, it is a liquid.

Various important temperatures may be identified between $T_m$ and $T_g$:

(a) The strain point. The glass has a viscosity of $3 \times 10^{14}$ P. Brittle fracture will precede plastic deformation below this temperature.

(b) The annealing point. The viscosity is $10^{13}$ P. The mobility of the molecules allows total relief of residual stresses to take place.

(c) The softening point. The viscosity is $4 \times 10^7$ P.

(d) The working point. The viscosity is $10^4$ P. The glass is easily deformed.

(e) The melting point. The viscosity is 100 P and the glass may be considered a liquid.

Glass forming operations are performed between the working point and the softening point. The raw materials are heated to a temperature where melting occurs and then allowed to cool. After production of the glass, the glass may be blown, pressed, drawn or fibre formed.

In pressing, a glass sheet is heated to the working temperature and then formed by pressing in a cast iron mould having the shape of the component to be made. Glass articles like rods, tubes sheets etc. on the other hand are made by drawing the molten glass through an orifice and then allowing the article to cool slowly. Glass fibers are formed in a similar process. Blowing, which may be manual or automated, is used to produce glass bottles, jars, etc. The process is similar to that for plastics outlined in chapter thirteen. The starting material is a solid piece, which is placed into a parison mould. It is pressed mechanically into a temporary shape. This is then placed into a finishing mould and air is blown into it. This forces the glass to take the shape of the mould.

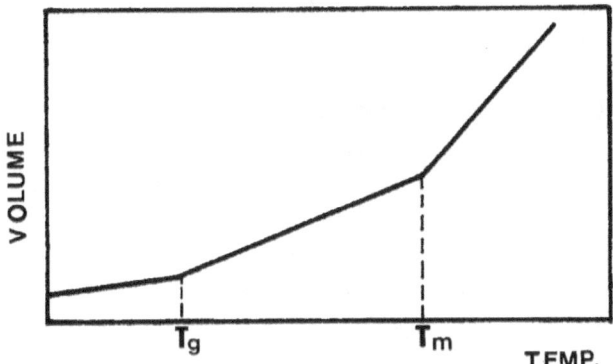

**Fig. 14.12** Variation of specific volume with temperature for a glass

## 14.8.2 Particulate Forming

The starting raw materials (usually clays) for this process are in the form of particles. The particle size is first reduced by mechanical grinding or milling. The ingredients are then thoroughly mixed and (where necessary) water is added. The clay then becomes plastic (the effect is termed hydroplasticity) and may be easily formed into the desired shape. Correct water ratio is essential if the product is to retain its shape throughout subsequent processing. The forming may be done by extrusion through a die opening having the shape desired. This method is used to produce clay bricks, pipes, tiles, etc.

Alternatively, slip casting may be used. Slip, which is a suspension of clay in water, is poured into a porous mould made of plaster of Paris. The mould absorbs water from the slip leaving a layer of clay on the mould wall. The process is repeated until the required wall thickness is achieved. Eventually, the mould is broken to release the finished product.

The hydroformed or slip cast article is then dried to remove excess water. The dried piece is termed a green. Shrinkage usually results during the drying process resulting in cracks unless the water removal rate is carefully controlled.

The dried article is then fired. Firing involves heating to a temperature at which some of the ingredients melt. The molten glass coats the surfaces of the rest of the particles thereby holding them together. The density and the strength of the article are increased, while the porosity is decreased. The process of formation of a liquid glass phase is termed vitrification and its degree determines the room temperature properties of the article.

In summary, processing of particulate materials which are usually clays involves grinding and milling, hydroplasticing, forming, drying and firing, or grinding and milling, slip formation, slip casting, drying, and firing.

### 14.8.3 Other Processing Methods

Ceramic abrasives may be bonded to grinding wheels by a glassy ceramic or an organic resin such that significant porosity is retained. Flow of air through the pores helps cool the wheel during grinding. Alternatively, the grinding particles may be coated onto a cloth or paper (sand paper) using the same methods, i.e., a glassy ceramic or a resin. Finally, ceramic grinders/ polishers may also be used as loose particles suspended in water or oil.

Electronic and magnetic ceramics may be processed by powder processing which is similar to the processes of powder metallurgy. A ceramic powder plus water binder is first pressed into the desired shape. The pressure application may be uniaxial or isostatic. Moreover, the process may be carried out in a cold or hot environment. The compacted mass takes the shape of the die and platens through which the pressure is applied.

The formed article is then fired whereupon sintering takes place. The sintering involves bulk diffusion of atoms between particles pressed together. This takes place below the melting temperature; hence no liquid phase is involved.

Finally, cements, plaster of Paris and lime (sometimes mixed with other ingredients e.g., sand) are processed through hydration. When these ceramics are mixed with water, a hydration reaction takes place which leads to setting. Hydration may take several months or even years to complete. It should be noted that strengthening here takes place following addition of water rather than removal of water (drying).

## PROBLEMS.

14.1 Calculate the minimum cation to anion ratio for a coordination number of four. (Ans: 0.225).

14.2 The cation to anion ratio in $CaCl$ is 0.732. Calculate its atomic packing factor. (Ans: 0.73)

14.3 Given that the crystal structure of $Al_2O_3$ consists of a HCP arrangement of anions while the cations occupy the octrahedral positions:

(a) Calculate the fraction of octrahedral positions occupied. (Ans: 66.7 %)

(b) Sketch a unit cell of $Al_2O_3$ showing the occupied octrahedral positions.

14.4 Determine the edge length of a unit cell of FeO given that FeO has a rock salt crystal structure and a density of 5.7 $kg/m^3$. (Ans: 0.434 nm)

14.5 Given that one polymorph of $SiO_2$ has a cubic crystal structure with an edge length of 0.7 nm and a density of 23.3 $kg/m^3$, calculate the number of $Si^{4+}$ and $O^{2-}$ ions per unit cell of $SiO_2$. (Ans: 8, 16)

# Further Reading

Van Vlank, L. H. Physical Ceramics for Engineers, Addison-Wesley Co. Reading, Ma. 1964.

Kingery, W. D. Introduction to Ceramics, John Wiley and Sons, New York, 1960.

Kingery, W. D. (Ed.), Ceramic Fabrication Processes, John Wiley and Sons, New York, 1958.

Rhines, F.N. Phase Diagrams in Metallurgy, McGraw-Hill, New York, 1956.

# CHAPTER FIFTEEN

# INTRODUCTION TO COMPOSITE MATERIALS

## 15.1  INTRODUCTION

In the broadest sense, a composite material may be defined as any material which is composed of two or more materials that are different, (physically, chemically or both) combined into one. If this definition is adopted, then virtually all materials in use are composites. As has been stated earlier, all metallic materials are either intentionally alloyed or contain impurities. Likewise, most plastic materials are either blended, or contain various additives to improve one property or another. These materials may be termed micro-composites or phases since the components in question are mixed at the micro scale and may only be distinguished under high magnification.

At the operational level (which is the position taken in this book), a composite material is defined as a material consisting of two or more physically distinct and mechanically separable materials (with an interface between them) that are combined to give performance in service that is superior in some specific respect to those of the individual components. In other words, the components are distinct at the macro scale. Good examples of composites that have been in use a long time are galvanized steel where the corrosion resistance of zinc and the strength of steel are combined to give a material that is both strong and corrosion resistant; and reinforced concrete where the good compressive strength of concrete and the good tensile strength of steel are combined.

The idea behind the development of composite materials comes from nature itself. Most materials existing naturally are indeed composites. The best examples are wood, bamboo, and bone. All of these are combinations of materials with different properties combined in specific ways to give specific combinations of seemingly contradictory properties. In bone, bamboo and wood for example, the composite ensures very high rigidity in the longitudinal direction (which is the direction of the highest loads to which the materials are subjected) compared to the other directions. These materials may therefore be classified as natural composites.

As a class of materials, composites are relatively new (compared to ceramics for example). However, their special advantages that will be discussed shortly, have led to their widespread use to an extent that one finds composites in nearly all fields of materials usage. Some of these are:

(a) Aircraft manufacture (wings, fuselage, etc.)

(b) Automobile manufacture (body parts, panels, bumpers, etc.)

(c) Boat manufacture (hulls, masts, etc.)

(d) Chemical industry (pressure vessels, containers, pipes, etc.)

(e) Household goods (furniture, ladders, etc.)

(f) Sports equipment (golf clubs, pole vault poles, tennis rackets, bicycle frames, fishing rods, canoes, skis, etc.)

(g) Electrical appliances (switchgear, panels, etc.)

One major limitation of most composite materials is their anisotropy. However, even this (anisotropy) may be used to advantage as will be explained presently.

## 15.2   NATURE OF COMPOSITES

As already stated, composites consist of two or more phases. One phase is discontinuous and is embedded in a continuous phase. The discontinuous phase provides reinforcement to the continuous phase, which is termed the matrix. The resulting properties of the composite depend on properties of the components, on their distribution (relative proportions, orientation, shape, and size of the reinforcing phase), and on the interaction between the components.

For the material to function as a composite, the discontinuous phase must carry a significant portion of the load. The matrix, while carrying a portion of the load, is primarily meant to:

1. Transfer the load to the discontinuous phase

2. Bind the discontinuous phase into a single unit

3. Protect the surfaces of the (usually delicate) discontinuous phase.

The two materials may act synergistically, i.e., their combined effect may exceed the sum of the effects of each acting individually.

In theory, any combination of the three primary classes of materials (metals, polymers and ceramics) may be used as either the matrix or the discontinuous phase. Thus, it would be possible to have metal matrix with metal reinforcement or a ceramic matrix with polymer as reinforcement. In practice however, only a few combinations (which will be enumerated shortly) have produced usable composites due to other factors like ease of fabrication, compatibility, desired properties, cost, etc.

## 15.3 CLASSIFICATION OF COMPOSITES

Several criteria may be used as the basis for classification of composites. Some of these are: whether the composite is natural or artificial (see section 15.1 above); the material class of the matrix. This leads to classifications like metal matrix composites or ceramic matrix composites; the nature (i.e., size, shape and distribution) of the discontinuous phase i.e., whether the reinforcement is fibrous or particulate, aligned or randomly distributed, etc. Most authors adopt the last of these, and using the same, composites may be classified as shown in figure 15.1.

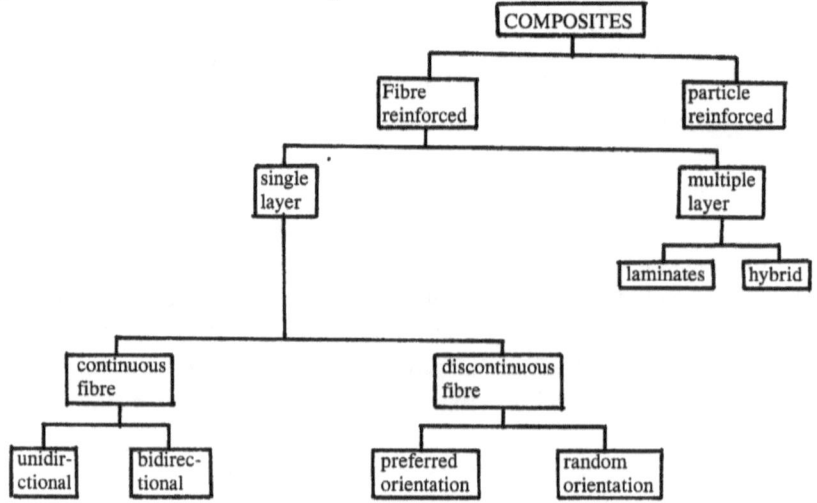

**Fig. 15.1** Classification of composites

Currently fibre reinforced composites comprise the largest class of composites in use and are hence the best studied. In most cases, the matrix is either a thermosetting resin or a thermoplastic (see chapter 13). The fibers are strong and stiff but usually brittle. The plastic is ductile and resistant to chemical attack thus providing protection to the fibers. The resulting material has strength and stiffness close to those of the fibre and a chemical resistance close to that of the plastic.

## 15.4 FIBRE REINFORCED COMPOSITES
### 15.4.1 Introduction
A fibre is a material whose dimensions in the longitudinal direction are much higher than in the other directions. For use in composites, inorganic

fibers are made mainly from glass, graphite (carbon) or boron. The fibers are specially manufactured to ensure that they have no internal flaws, which may act as stress raisers. Their diameters are about the size of a single crystal. Moreover, the crystals may be oriented such that the crystal direction in which the crystal is strongest is in the longitudinal direction of the fibre. The resulting fibers have exceptionally high strength and stiffness in the longitudinal direction. However, some fibers may also be organic in nature. In this case, the molecules are oriented in the longitudinal direction. The most common organic fibre is amarid polymer, known by the trade name, Kevlar.

The fibers may themselves be classified as being long or short. Fibre reinforced composites with long fibers are termed continuous fibre reinforced composites (CFRC), while those with short fibers are termed discontinuous fibre reinforced composites (DFRC). If the fibers are oriented in one direction, the resulting composite is defined as unidirectional. Unidirectional fibre reinforced composites are anisotropic having high strength and stiffness in the direction of the fibers but low strength/stiffness values in the transverse direction. To produce usable composites from unidirectional composites, layers of the later are stack together such that each layer is oriented in a different direction. The resulting material is termed a laminate. Laminates can be tailored to have controlled anisotropy thus allowing for optimum use of material. As an example, if the desired component has higher stresses in a known direction, the fibers can be preferentially aligned in this direction. In this manner, attributes like high strength to weight ratio may be achieved. A good example is a cylindrical pressure vessel. Here, the stresses in the circumferential direction are twice those in the longitudinal direction. A tailored design will be such that there are twice as many fibers in the circumferential direction compared to the longitudinal direction.

In discontinuous fibre composites, it is not possible to control the orientation of the fibers. They may therefore be considered as randomly oriented. The resulting composite in this case is quasi isotropic.

## 15.4.2 Unidirectional Continuous FRC

A schematic representation of a unidirectional continuous fibre reinforced composite is shown in figure 15.2. The fibers extend from one end of the composite to the next in the longitudinal direction. In this case, the load may be considered to be carried directly by the fibers thus reducing the role of the matrix to that of protection and binding of the fibers.

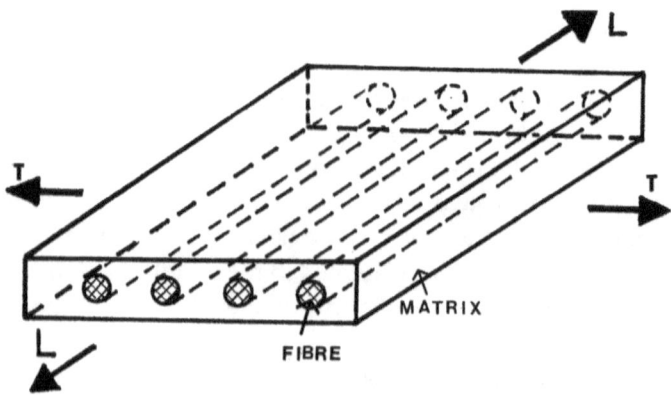

**Fig. 15.2** Schematic of a unidirectional continuous FRC

It is important to be able to predict the properties of a composite from the known properties of the fibre and matrix. This requires that the proportions of the components be determined. The most common method of expressing this is to define the volume fraction of each phase. First, the following definitions are made:

$v_f$   = volume of fibers in the composite
$v_m$  = volume of the matrix in the composite
$v_c$   = volume of the composite

Then,

$$v_c = v_f + v_m \tag{15.1}$$

The volume fraction of fibre, $V_f$, can then be defined as:

$$V_f = \frac{v_f}{v_c} \tag{15.2}$$

Likewise, the volume fraction of matrix is given by:

$$V_m = \frac{v_m}{v_c} \tag{15.3}$$

From the three equations above, it is clear that:

$$V_m = 1 - V_f \tag{15.4}$$

The relative quantities may also be expressed in weight percent or as weight fractions. Thus, the weight fraction of fibre, $W_f$ is given by:

$$W_f = \frac{w_f}{w_c} \tag{15.5}$$

where $w_f$ and $w_c$ are the respective weights of the fibre and composite. Then, if the weight of the matrix is designated $w_m$:

$$w_c = w_f + w_m$$
$$V_c\rho_c = V_f\rho_f + V_m\rho_m$$

$$\rho_c = \frac{v_f}{v_c}\rho_f + \frac{v_m}{v_c}\rho_m$$

$$\rho_c = V_f\rho_f + V_m\rho_m \qquad (15.6)$$

where $\rho_f$, $\rho_m$, and $\rho_c$ are the respective densities of the fibre, matrix and composite.

The relation between volume fraction and weight fraction can be easily determined:

$$V_f = \frac{v_f}{v_c} = \frac{m_f}{m_c}\frac{\rho_c}{\rho_f}$$

$$V_f = \frac{\rho_c}{\rho_f}W_f \qquad (15.7)$$

Similarly:

$$V_m = \frac{\rho_c}{\rho_m}W_m \qquad (15.8)$$

The calculated and actual densities may differ due to the presence of voids in actual composites.

Stress sharing in UCFRC (longitudinal direction)

The strength of the composite depends on the strength of the constituents, their distribution and the nature of the chemical and physical interaction between them. To a first approximation, the following assumptions may be made while predicting the strength of UCFRC:

(a) The fibers are continuous and aligned i.e., they run from one end of the composite to the next, are uniform in geometry and distribution and are laid parallel to each other.

(b) There is a perfect bond between the fibre and matrix.

(c) A unidirectional load is applied parallel to the fibers.

Under these conditions, the strain in the matrix, fibre and composite are all equal to each other (so called iso-strain condition). On the other hand, the load carried by the composite, $P_c$, is shared by the matrix ($P_m$) and the fibre ($P_f$). Hence:

$$P_c = P_f + P_m$$
$$\sigma_c A_c = \sigma_f A_f + \sigma_m A_m$$

$$\sigma_c = \sigma_f \frac{A_f}{A_c} + \sigma_m \frac{A_m}{A_c} \tag{15.9}$$

where, $A_f$, $A_m$ and $A_c$ are the respective total cross section areas of the fibre, matrix and composite, while $\sigma_f$, $\sigma_m$ and $\sigma_c$ are the stresses in the respective phases. From the assumptions already made:

$$\frac{A_f}{A_c} = \frac{v_f}{v_c} = V_f$$

$$\sigma_c = \sigma_f V_f + \sigma_m V_m \tag{15.10}$$

Stiffness of UCFRC (longitudinal direction)
Let the strain (which is equal in all three phases) change by an infinitesimal amount, $\delta\varepsilon$. Let the corresponding increase of stresses be $\delta\sigma_c$, etc. Then from equation 15.10:

$$\frac{d\sigma_c}{d\varepsilon} = \frac{d\sigma_f}{d\varepsilon} V_f + \frac{d\sigma_m}{d\varepsilon} V_m$$

If all the materials are in the elastic range, then $\delta\sigma / \delta\varepsilon = E$. Hence:

$$E_c = E_f V_f + E_m V_m \tag{15.11}$$

Equations 15.10 and 15.11 state that the contribution of each phase to the composites' stress or stiffness is proportional to their volume fraction. This is termed the rule of mixtures. These equations are applicable only for tensile loading. In compression, the fibers offer no strengthening as they fail by buckling. A composite's compressive properties are therefore the same as those of the matrix.

Load sharing in UCFRC (longitudinal)
Since the strains are all equal, and if the assumption of linear elasticity still holds, then:

$$\varepsilon_f = \frac{\sigma_f}{E_f} = \varepsilon_m = \frac{\sigma_m}{E_m}$$

$$\frac{\sigma_f}{\sigma_m} = \frac{E_f}{E_m} \qquad\qquad (15.12)$$

Similarly:

$$\frac{\sigma_f}{\sigma_c} = \frac{E_f}{E_c} \qquad\qquad (15.13)$$

Furthermore,

$$\frac{P_f}{P_m} = \frac{\sigma_f \cdot A_f}{\sigma_m \cdot A_m} = \frac{\sigma_f(A_m / A_c)}{\sigma_m(A_m / A_c)} = \frac{\sigma_f V_f}{\sigma_m V_m}$$

Therefore:

$$\frac{P_f}{P_m} = \frac{E_f}{E_m} \cdot \frac{V_f}{V_m} \qquad\qquad (15.14)$$

and:

$$\frac{P_f}{P_c} = \frac{P_f}{P_f + P_m} = \frac{E_f \cdot V_f}{E_f V_f + E_f V_m} \qquad\qquad (15.15)$$

Equations 15.14 and 15.15 illustrate that maximum benefit is achieved when the ratio $E_f / E_m$ is highest.

Example 15.1: A unidirectional continuous fibre reinforced composite consists of 45 % glass fibers (E = 70 GPa) by volume. The fibre are embedded in polyester resin (E = 3.5 GPa). Calculate:
(a) The modulus of elasticity in the longitudinal direction.
(b) The load carried by each phase when a load of 12 kN is applied to the composite.
(c) The corresponding strain if the cross section area of the composite is 260 mm². Assume linear elastic conditions.
(d) The corresponding stresses in the fibre and matrix.

Answer:

(a) $V_f = 0.45$ and $V_m = 1 - V_f = 0.55$

Substituting these values into equation 15.10:

$E_c = 33.4$ GPa

(b) Substituting the values of $V_f$, $V_m$, $E_f$ and $E_m$ into equation 15.15, we get:

$$\frac{P_f}{P_c} = \frac{70x0.45}{70x0.45 + 3.5x0.55} = 0.94$$

$\Rightarrow P_f = 11.3$ kN

and:  $P_m = P_c - P_f = 0.7$ kN

It is noted that the fibers carry the bulk of the load.

(c) The strain is the same in all three phases. Hence:

$\varepsilon = \varepsilon_c = \sigma_c / E_c = P_c / (A_c E_c)$

$$- = \frac{12x10^3}{26x10^{-5} x33.4x10^9} = 1.38x10^{-3}$$

(d) Noting that $V_f = A_f / A_c$,

$$\sigma_f = \frac{P_f}{A_f} = \frac{P_f}{A_c V_f} = \frac{11,300}{260x10^{-6}x0.45} = 96.6\_MPa$$

In a similar manner,

$\sigma_m = 4.9$ MPa

## Critical and Minimum Fibre Volume Fractions

If the fibers are to provide reinforcement, the ultimate tensile strength of the composite must be higher than that of the unreinforced matrix. Failure occurs first in the fibers since they have a lower fracture strain compared to the matrix. From equation 15.10:

$\sigma_{cu} = \sigma_{fu}V_f + \sigma_m'(1 - V_f)$  (15.16)

where, $\sigma_{cu}$, $\sigma_{fu}$ are the ultimate tensile strengths of the composite and fibre respectively, while $\sigma_m'$ is the stress in the matrix at the time of fibre fracture. If the ultimate tensile strength of the matrix is $\sigma_{mu}$, then for effective reinforcement:

$$\sigma_{fu}V_f + \sigma_m'(1-V_f) > \sigma_{mu} \qquad (15.17)$$

The least fibre volume fraction that satisfies inequality 15.17 is the critical volume fraction, $V_{crit}$. i.e.:

$$V_{crit} = \frac{\sigma_{mu} - \sigma_{m'}}{\sigma_{fu} - \sigma_{m'}} \qquad (15.18)$$

If $V_f$ is less than $V_{crit}$, the fibers stretch to failure immediately on application of the load and the matrix has to carry the whole load. Equation 15.9 is therefore not valid for calculation of the ultimate tensile strength which must then be calculated from:

$$\sigma_{cu} = \sigma_{mu}V_m = (1 - V_f)\,\sigma_{mu} \qquad (15.19)$$

Equation 15.19 shows that for $V_f$ less than $V_{crit}$, $\sigma_{cu}$ decreases with $V_f$. Equations 15.9 and 15.19 are plotted in figure 15.3 as functions of $V_f$.

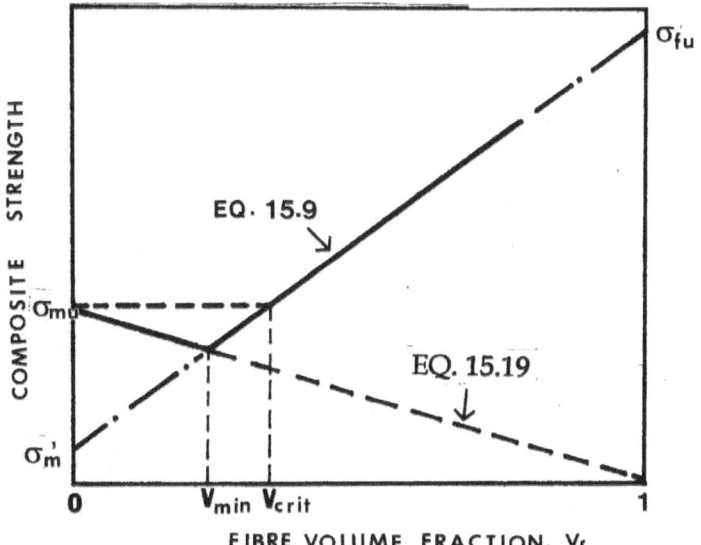

**Fig. 15.3** Variation of composite strength with fibre volume fraction

The volume fraction corresponding to the minimum composite strength is designated the minimum volume fraction, $V_{min}$, and may be calculated by equating the composite strength calculated from equation 15.9 to that calculated from equation 15.19, i.e.:

$$\sigma_{fu}V_{min} + \sigma_m' (1 - V_{min}) = \sigma_{mu} (1 - V_{min})$$

$$V_{min} = \frac{\sigma_{mu} - \sigma_m'}{\sigma_{fu} + \sigma_{mu} - \sigma_m'} \tag{15.20}$$

If the fibers have some ductility, the graphs shown in figure 15.3 have to be modified as shown with dotted line in the figure. In this case, no drop of composite tensile strength with volume fraction occurs since the fibers yield (but are still able to carry load) instead of fracturing.

Although the above equations do not suggest it, there is a limiting maximum value of volume fraction allowable. Beyond this value, the fibers come into contact with each other and the protection and binding functions of the matrix are lost. The fibers then fracture prematurely and a drop in composite strength results. This limiting volume fraction is about 0.75 for most brittle fibers.

After fibre fracture, there is a sudden drop in load, but the matrix may continue deforming plastically. Moreover, some strain hardening may take place in the matrix. Assuming a polymeric matrix, the resulting stress strain curve will have the form shown in figure 15.4.

Example 15.2: A UCFRC with the following properties is to be produced:

| | |
|---|---|
| Fibre volume fraction: | 0.4 |
| Longitudinal modulus of elasticity: | 60 GPa |
| Longitudinal tensile strength: | 1.1 GPa |
| Epoxy matrix: | $E = 3.2$ GPa; $\sigma_{mu} = 70$ MPa. |

If the following fibres are available: Carbon ($E = 250$ GPa, $\sigma_{fu} = 3.2$ GPa); Glass ($E = 70$ GPa, $\sigma_{fu} = 3.5$ GPa; Kevlar 49 ($E = 150$ GPa, $\sigma_{fu} = 2.5$ GPa), determine, with reasons, which fibre(s) may be used.

Answer:

Consider carbon: $E_c = V_fE_f + V_mE_m = 0.4 \times 250 + 0.6 \times 3.2$ GPa
$$= 102 \text{ GPa}$$
$$\sigma_{cu} = V_f\sigma_{fu} + V_m\sigma_{mu} = 0.4 \times 3200 + 0.6 \times 70 \text{ MPa}$$
$$= 1.32 \text{ GPa}$$

Therefore, carbon fibre may be used.

Consider glass fibre: $E_c = 0.4 \times 70 + 0.6 \times 3.2$ GPa
$$= 29.9 \text{ GPa}$$

Hence glass fibre would not meet the requirements.

Consider Kevlar 49: $E_c = 0.4 \times 150 + 0.6 \times 3.2$ GPa
$$= 61.9 \text{ GPa}$$

$$\sigma_{cu} = 0.4 \times 2500 + 0.6 \times 70 \text{ MPa}$$
$$= 1.04 \text{ GPa}$$

This is less than the tensile strength required. Hence only the carbon fibre may be used for this application.

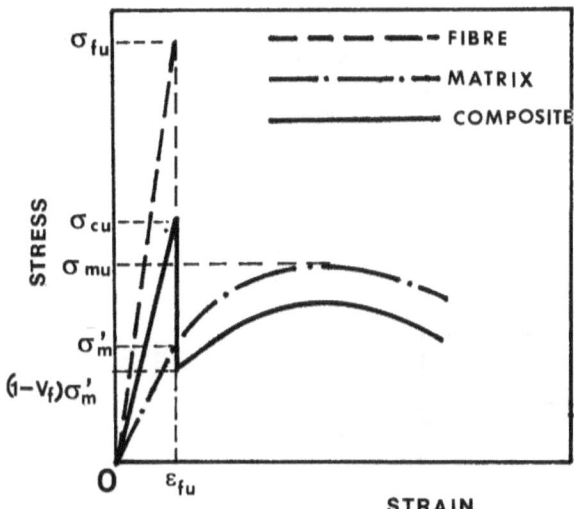

**Fig. 15.4** Idealized longitudinal stress/strain curves for a UCFRC:

Note that in calculating the composite strength, $\sigma_{mu}$ is used in place of $\sigma_m'$. The reader should verify that the error introduced by this assumption is negligible by recalculating the values assuming that $\sigma_m' = {}^1/_2\sigma_{mu}$. This is because the contribution of the matrix strength to the composite strength is minimal.

### Transverse Stiffness/strength in UCFRC
The assumptions made earlier with regard to longitudinal properties apply for transverse loading too. A schematic showing transverse loading of UCFRC is shown in figure 15.5.

From the assumption of uniform distribution it may be inferred that the total thicknesses of both matrix and fibre are proportional to their respective volume fractions. Moreover, the total elongation of the composite is the sum of the individual elongations of the fibre and matrix, i.e:

$$\delta_c = \delta_f + \delta_m \tag{15.21}$$
$$\Rightarrow \quad \varepsilon_c t_c = \varepsilon_f t_f + \varepsilon_m t_m$$

$$\Rightarrow \qquad \varepsilon_c = \varepsilon_f V_f + \varepsilon_m V_m \qquad (15.22)$$

Assuming linear elasticity in all phases, and noting that $\varepsilon = {}^{\sigma}/_E$:

$$\frac{\sigma_c}{E_c} = \frac{\sigma_f V_f}{E_f} + \frac{\sigma_m V_m}{E_m} \qquad (15.23)$$

The stress across each component is the same (iso-stress condition), i.e., $\sigma_c = \sigma_f = \sigma_m$, hence equation 15.23 becomes:

$$\frac{1}{E_c} = \frac{V_f}{E_f} + \frac{V_m}{E_m} \qquad (15.24)$$

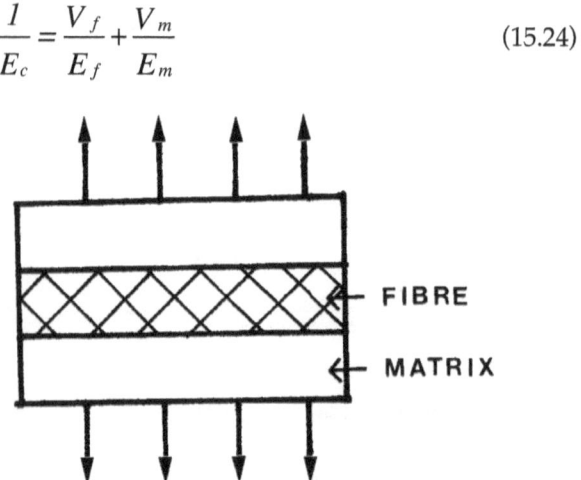

**Fig. 15.5** Schematic showing transverse loading of UCFRC

The variation of $E_c/E_m$ with the fibre volume fraction is shown schematically in figure 15.6 for both longitudinal and transverse loading. From the figure, it is evident that the fibres are not very effective in increasing the stiffness in the transverse direction except at values of volume fraction approaching one.

Equation 15.24 is only approximate since the dispersion of fibres is not exactly uniform and hence when a plane is traversed, it may happen that both fibre and matrix are traversed. This would lead to a violation of the iso-stress assumption. Moreover, the strains in the fibre and matrix are different, leading to a strain mismatch at the interface. Lastly, there is a Poisson's ratio mismatch, which sets up longitudinal stresses.

Semi empirical relationships that give better agreement with experimental results have been developed (see further reading list at the end of the chapter). One example is that developed by Halpin and Tsai:

$$E_c = \frac{1+\xi\eta V_f}{1-\eta V_f} E_m \qquad (15.25)$$

where $\xi$ is a constant approximately $=2$, and

$$\eta = \frac{(E_f/E_m)-1}{(E_f/E_m)+\xi} \qquad (15.26)$$

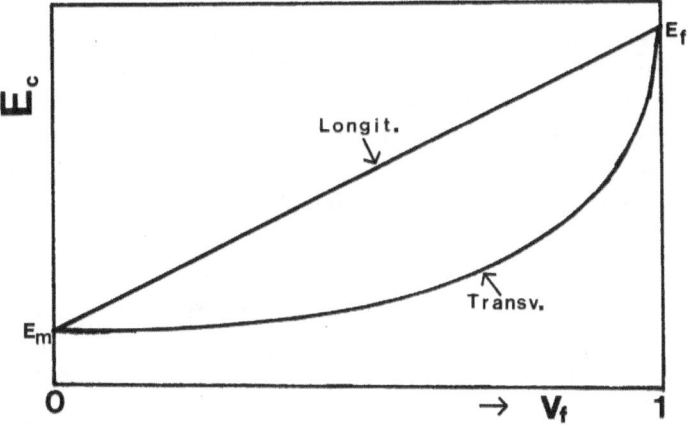

**Fig. 15.6** Variation of composite modulus with fibre volume fraction

Since there is no load sharing in the transverse direction, the strength of the composite in this direction is lower than the matrix strength, an effect compounded by the differences in stiffness between the two phases. The later leads to a constraint in extension of the matrix, which in turn leads to strain concentration at the fibre matrix interface. The net result is failure at lower loads i.e., a reduced strength.

Example 15.3: For the composite of example 15.1, calculate the modulus in the transverse direction.
Answer: $V_f = 0.45$; $E_f = 70$ GPa; $E_m = 3.5$ GPa
Using equation 15.23:

$$\frac{1}{E_c} = \frac{0.45}{70} + \frac{0.55}{3.5}$$

From which: $\qquad E_c = 6.1$ GPa

### 15.4.3 Discontinuous Fibre Reinforced Composites (DFRC)

In DFRC, the fibres do not run from one end of the composite to the other, i.e., the fibre is shorter than the composite. For this reason, they are also termed short fibre composites. The short fibres may be aligned (i.e., all fibres have their longitudinal axes principally in the longitudinal direction of the composite) or randomly oriented. The composites in which the fibres are randomly oriented have the advantage of isotropy and ease of moulding.

Since the fibres do not run from end to end, the load in this case is applied to the matrix, which transfers it (the load), via the interface, to the fibres. As will be shown presently, the ends of the fibres do not transfer as much load as the middle part. This effect is more pronounced the shorter the fibres are. For this reason, the length of the fibres is an important parameter in the analysis of short fibre composites.

When a load is applied to the matrix, the matrix attempts to flow past the fibre. This sets up a shear stress, $\tau$, at the fibre-matrix interface. There is also an effect due to friction between the phases but this contribution is minimal and will not be considered here.

Consider a short aligned fibre embedded in a matrix as shown in figure 15.7, and the infinitesimal element shown alongside:

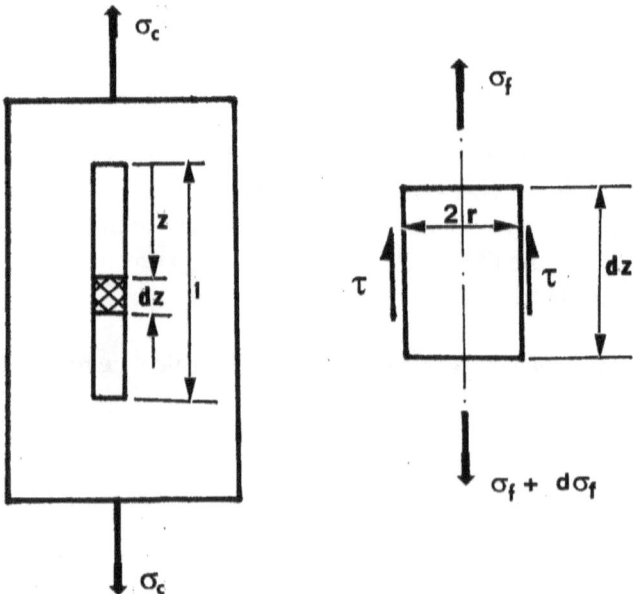

**Fig. 15.7** Mechanics of a short fibre composite

Considering the equilibrium of the element:

$$\Sigma F(z) = 0 \ \Rightarrow \ \pi r^2 \sigma_f + 2\pi r \tau dz = (\sigma_f + \delta \sigma_f)\pi r^2$$

where r is the radius of the fibre and $\sigma_f$ is the stress in the fibre at some distance, z from one end. Then:

$$d\,\sigma_f = \frac{2\tau}{r}\,dz$$

$$\Rightarrow \sigma_f = \frac{2}{r}\int_0^z \tau dz$$

At $z = 0$, $\sigma_f = 0$, i.e., there is no tensile stress at the fibre ends. Treating the matrix as a rigid-perfectly plastic material, it may be assumed that $\tau = \tau_{mY} =$ constant, where $\tau_{mY}$ is the ultimate strength of the matrix in shear or the interfacial shear strength whichever is smaller. Then:

$$\sigma_f = \frac{2\tau_{mY}\,z}{r} \tag{15.27}$$

Equation 15.27 is applicable only where $z < 1/2$, since at $z = 1$, the stress must again be zero. As a consequence, the maximum stress occurs at $z = 1/2$. Hence:

$$\sigma_{fmax} = \frac{\tau_{mY}\,l}{r} \tag{15.28}$$

To utilise the full capacity of the fibre, $\sigma_{fmax}$ should be equal to the fibre ultimate tensile strength. This means that a certain length, $l_c$, is required before full capacity of the fibre can be realized. This value of l, termed the critical length, $l_c$, can be calculated by setting $\sigma_{fmax}$ equal to $\sigma_{fu}$ in equation 15.28. From this:

$$l_c = \frac{\sigma_{fu}\,r}{\tau_{mY}} \tag{15.29}$$

When the fibre length is less than the critical length, fibre fracture does not take place. Instead, failure occurs at the matrix fibre interface and the fibres are pulled out (so called fibre pull out). If the composite is loaded in bending, failure occurs by splitting at the interface. This provides a simple experimental method for determining the critical length. Fibre pull out tests are carried out for fibres of different lengths as represented schematically in figure 15.8 The largest value of l for which fibre fracture occurs instead of fibre pull out is an estimation of the critical length.

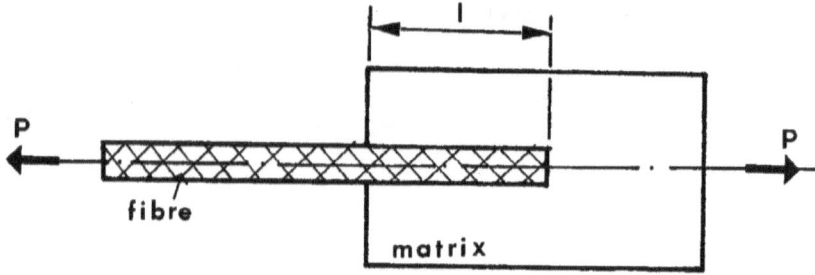

**Fig. 15.8**   Schematic representation of the fibre pull-out test

The ratio $^1/_d$ (or $^1/_{2r}$) is termed the fibre aspect ratio. The value of this corresponding to the critical length is the critical aspect ratio. The variation of fibre stress with the distance along the fibre, z, is shown in figure 15.9 for the three cases: $l < l_c$, $l = l_c$, and $l > l_c$.

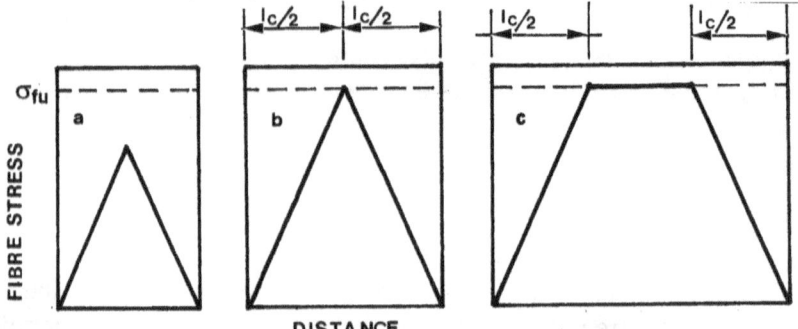

**Fig. 15.9**: Variation of fibre stress with axial distance for fibres of different lengths.

Since the ends of the fibres are not stressed to the full capacity, the efficiency of strengthening is reduced. The result is that both Modulus of elasticity and composite tensile strength in short fibre composites are less than the corresponding values calculated for continuous fibre composites. It is possible to define an average fibre stress, $\sigma_{fav}$ as:

$$\sigma_{fav} = \frac{1}{l} \int_0^l \sigma_f \, dz$$

$$\sigma_{fav} = \frac{1}{2}\sigma_{fmax} = \frac{\tau_{mY}\, l}{2r} ; l < l_c \qquad (15.30)$$

$$\sigma_{fav} = \sigma_{fmax}\left(1 - \frac{l_c}{2l}\right) = \frac{\tau_{mY}\, l}{r}\left(1 - \frac{l_c}{2l}\right); l > l_c \qquad (15.31)$$

It should be noted that more rigorous treatments of the mechanics of short fibre composites (e.g., those by Cox, Outwater, Kelly and Tyson, etc.) are available and the interested reader is referred to the further reading list at the end of the chapter. The simplified treatment presented here will suffice except where greater accuracy requirements justify the extra complexity.

Stiffness of short fibre composites

When the length of the fibres is much greater than the critical length, (approximately 15 times) the end effects may be neglected and the fibres treated as continuous fibres. The modulus of elasticity may then be calculated from the equations developed earlier. When $l_c < l < 15\, l_c$, the rule of mixtures is modified by application of a length correction factor, $\eta_l$. For longitudinal loading, the modulus becomes:

$$E_c = \eta_l E_f V_f + E_m V_m \qquad (15.32)$$

For carbon and glass fibres in epoxy resin, $\eta_l$ varies from 0.2 to 0.9 for fibre lengths between 0.1 mm and 1.0 mm. Alternatively, semi empirical formulae developed from finite element analysis may be found in the literature. One example is that due to Haplin and Tsai (see reading list at the end of the chapter):

$$E_c = \left[\frac{1 + \frac{l}{r}\eta V_f}{1 - \eta V_f}\right] E_m$$

where,

$$\eta = \frac{E_f / E_m - 1}{E_f / E_m + \frac{l}{r}} \qquad (15.34)$$

If the orientation of the fibres is random, the following empirical equation (see further reading list at the end of the chapter) may be used to estimate the composite modulus:

$$E_c = \frac{3}{8}E_{CL} + \frac{5}{8}E_{CT} \qquad (15.35)$$

where $E_{CL}$ and $E_{CT}$ are the composite moduli in the longitudinal and transverse directions, respectively for an aligned short fibre composite in which the fibres have the same aspect ratio and volume fraction as the randomly oriented composite under consideration.

Stress partitioning in SFRC

The stress in aligned short fibre composites can be estimated from a modification of equation 15.6 as:

$$\sigma_c = \sigma_{fav}V_f + \sigma_{m'}V_m \qquad (15.36)$$

where $\sigma_{fav}$ is the average fibre stress as defined earlier. Similarly, the composite ultimate tensile strength can be calculated from a modification of equation 15.11. Here, three cases arise. In the first case, $l < l_c$. The fibres do not fracture and composite strength depends on the matrix or interface failure. Combining equations 15.30 and 15.36:

$$\sigma_{cu} = \frac{\tau_{mY}l}{2r}V_f + \sigma_{m'}V_m \; ; l < l_c \qquad (15.37)$$

When $l > l_c$, equations 15.31 and 15.36 are combined to give:

$$\sigma_{cu} = \sigma_{fu}(1 - \frac{l_c}{2l})V_f + \sigma_{m'}V_m \qquad (15.38)$$

$$\sigma_{cu} = \frac{\tau_{mY}l_cV_f}{r}(1 - \frac{l_c}{2l}) + \sigma_{m'}V_m \qquad (15.39)$$

Lastly, when $l \gg l_c$, the composite may be treated as continuous. Hence:

$$\sigma_{cu} = \sigma_{fu}V_f + \sigma_{m'}V_m \qquad (15.40)$$

As illustrated in example 15.2, $\sigma_{mu}$ may be used in place of $\sigma_{m'}$ without causing appreciable error.

Example 15.4: Calculate the longitudinal strength of a discontinuous aligned fibre composite given the following: Fibre: E = 70 GPa; $\sigma_{fu}$ = 3.6 MPa; l = 2.5 mm; d = 0.014 mm; $V_f$ = 0.28; Matrix: E = 2.4 GPa; stress in matrix at composite fracture = 6.3 MPa. Fibre-matrix bond strength: 108 MPa.

Answer:

First, the critical length is calculated:

$$l_c = \frac{\sigma_{fu}r}{\tau_{mY}} = \frac{3600x0.014}{108x2} = 0.23 \, mm$$

From which it is noted that l > $l_c$. Equation 15.38 is then used to calculate the composite strength. Substituting the relevant values:

$$\sigma_{cu} = 3600x0.28(1 - \frac{0.23}{2x2.5}) + 6.3x0.72 MPa = 970 MPa$$

Example 15.5: Determine the longitudinal modulus of elasticity of the composite in example 15.4.

Answer:

By application of equation 15.32, and taking $\eta_l = 0.9$

$E_c = 0.9 \times 70 \times 0.28 + 2.4 \times 0.72 = 19$ GPa

Alternatively, using the semi empirical equations:

$$\eta = \frac{E_f/E_m - 1}{E_f/E_m + l/r} = \frac{70/2.4 - 1}{70/2.4 + 2.5/0.007} = 0.073$$

Then from equation 15.33:

$$E_c = 2.4 \left[ \frac{1 + 375x0.073x0.28}{1 - 0.073x0.28} \right] = 20 \_ GPa$$

It is noted that the results obtained by the two methods agree reasonably well.

Short fibre composites in which the orientation is random are largely isotropic. The properties of these can be estimated from a modification of the rule of mixtures by the introduction of an efficiency factor, K. Then:

$$E_c = KE_fV_f + E_mV_m \qquad (15.41)$$

The empirically determined values of K is about $3/8$ for fibres distributed within a plane and $1/5$ for fibres distributed within a three dimensional space.

## 15.5   PARTICLE REINFORCED COMPOSITES
### 15.1   Large Particle Composites

As opposed to fibres, particles are approximately equiaxed. Effective reinforcement is achieved only if the particles are small and evenly distributed. It has been determined experimentally that the elastic properties of particle reinforced composites fall in between an upper bound predicted by the iso-strain assumption i.e:

$$E_c = V_p E_p + E_m V_m \qquad (15.42)$$

and a lower bound predicted by the iso-stress assumption i.e:

$$E_c = \frac{E_p E_m}{V_m E_m + V_p E_p} \qquad (15.43)$$

where, $E_p$ and $V_p$ are, respectively, the values of the modulus and volume fraction of the particles. Particle reinforced composites are largely isotropic. The increase in stiffness results in this case from the particles' ability to resist deformation. Increase in strength due to particle reinforcement is however, minimal.

The particles used may be metallic, polymeric or ceramic. Ceramic particle metal matrix composites are a good example of particle reinforced composites, and are termed cermets. They have been used for a long time in the manufacture of metal cutting tools. The particles in this case are mainly tungsten carbide (WC) or titanium carbide (TiC) embedded in a metal matrix.

Other common examples of particle reinforced composites are: (1) Rubber for tyre manufacture that contain up to 30 % of carbon black to improve strength, toughness, wear and abrasion resistance. (2) Concrete which is a ceramic-ceramic composite used extensively in building and other civil engineering works. In this case, cement is the matrix while sand (fine aggregates) and gravel (coarse aggregates) are the reinforcing phases.

## 15.5.2   Dispersion Strengthened Composites

These are similar to large particle composites except that the reinforcing particles are much finer. In general, the matrix is a metallic alloy while the particles are ceramic (ordinarily hard, inert oxides). Strengthening is achieved through interaction of the particles with dislocations in much the same way as obtains in precipitation hardening (see chapter seven). Though dispersoids are not as effective in strengthening as precipitates, these composites can retain their properties at high temperatures better. Common examples of dispersion strengthened composites are aluminium-alumina and nickel-thoria ($ThO_2$) systems.

## 15.6   STRUCTURAL COMPOSITES

These include laminates and hybrid composites. Their properties depend on both the properties of the constituents and on the design.

Laminates are two-dimensional sheets of unidirectional composites that are stacked together such that the orientation of the high strength direction varies from layer to layer. A common example of a laminate is plywood.

Sandwich or hybrid panels consist of two outer sheets enclosing a less dense core. The outer sheets, which may themselves be unidirectional composites, carry the bulk of the load, while the core resists deformation perpendicular to the plane.

Laminate theory forms an interesting part of the study of composite materials but this will not be considered further here. The interested reader is referred to the reading list at the end of the chapter.

## 15.7   FAILURE MODES IN COMPOSITES

Composites subjected to tensile loads may fail by a variety of mechanisms which include fibre fracture, matrix microcracking, debonding of the fibre from the matrix, interface shear failure, or delamination (in laminates). It is also possible to have mixed mode failure e.g., where cracks forming at different planes in the matrix are joined by debonding.

When loaded in compression, the most common failure mode is micro buckling of the fibres. Alternatively, failure may initiate by transverse splitting. Stresses are induced in the transverse direction due to Poisson's effect, which leads to interface failure. Lastly, compression failure may initiate from shear failure of the matrix.

## 15.8   PROCESSING OF COMPOSITES

Only a brief mention of the methods of producing composite materials is given here. The methods may be classified into two broad categories: open mould processes and closed mould processes. The open mould processes are relatively low cost since the cost of producing the mould is low.

Moreover, they may be used for production of large components. However, they suffer from the disadvantage that only one side of the component can attain a smooth finish. Closed mould processes on the other hand involve two part moulds, and hence may produce components with both surfaces smooth finished. Moreover, high production rates may be realized allowing for mass production. The matrix may be either a thermoplast or a thermoset, but the fibres are generally short fibres.

## 15.8.1   Open Mould Methods

These include hand lay up or contact moulding, spray up, vacuum moulding, and filament winding. Contact moulding or hand lay up involves placing chopped strand mats or woven mats rovings into a mould having the shape of the component to be produced. The matrix, which is usually a resin, is impregnated into the mould by successive painting or rolling until all air has been expelled, and the desired thickness achieved. The component is left to cure without addition of heat or pressure. In the spray up method, chopped fibre and resin are simultaneously sprayed into a mould and then rolled. The vacuum bag method on the other hand involves impregnating layers of fibres with resin and leaving the same to cure partially. What results is termed a pre-preg and for this reason, this method is also called the pre-preg method. The layers of pre-preg are stacked in the desired sequence on the mould surface. They are then covered with a rubber bag and consolidated using either pressure or a vacuum in an autoclave where the curing process is completed.

In filament winding, strands of fibre are passed over a bath of resin before winding in a winding machine onto a mandrel. The filaments are wound at predetermined angles and left until completely cured. Components made by this method include tubes, pipes, pressure tanks, etc. Finally, centrifugal casting involves introducing a mixture of resin and fibres into a rotating mould. The resin is left to cure on site.

## 15.8.2  Closed Mould Processes

These include compression moulding, injection moulding, transfer moulding, and pultrusion. The first three are similar to those already considered while discussing manufacture of plastics (chapter 13). The only difference is that in the case of composites, the charge is a mixture of short fibres and the plastic, or comes in the form of pre-preg sheets. Pultrusion (the composite equivalent of extrusion) on the other hand is used for components having a constant cross section e.g., tubes, rods, beams, etc.

Fibre rovings are impregnated with resin and pulled through dies having the desired cross section. These are then passed through curing dies, also having the same shape but heated to initiate or complete curing.

## PROBLEMS

15.1 Determine the percentage of the load carried by the fibres for the following continuous fibre composites:

(a) Glass fibre (E = 70 GPa), epoxy (E = 3.5 GPa) at a fibre volume fraction of 10 %. (Ans: 69 %).

(b) Carbon fibre (E = 430 GPa), epoxy (E = 3.5 GPa) at a fibre volume fraction of 50 %. (Ans: 99.2 %).

15.2 Calculate the ratios of fibre stress to matrix stress, and fibre stress to composite stress for the following longitudinally loaded CFRC: $V_f = 25$ %; $E_f = 400$ GPa; $E_m = 3.2$ GPa. (Ans: 1:125; 1:3.9).

15.3 The idealized stress strain curves of two matrix materials are shown below. Both matrices are reinforced with glass fibre ($V_f = 0.45$; $E_f = 70$ GPa; $\sigma_{fu} = 3$ GPa). If the resulting composites are longitudinally loaded, calculate the following: (a) the minimum and critical fibre volume fractions. (Ans: A: 0.0080, 0.0082; B: 0.0, 0.0). (b) The composite strengths at 1% and 4 % strains. (Ans. A: 354 MPa, 1299 MPa. B: 354 MPa, 1279 MPa). Note: assume the fibres remain linear elastic up to fracture.

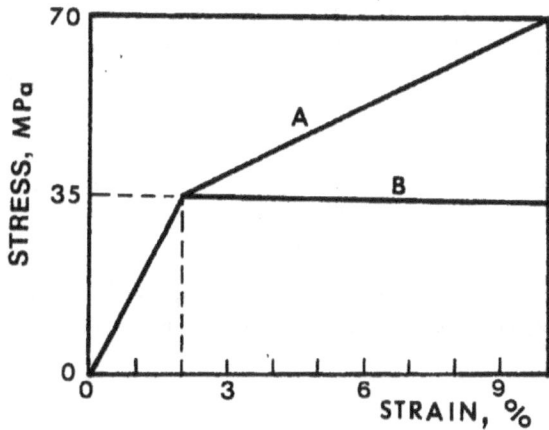

**Fig. Q. 15.3**

15.4 Calculate the modulus of elasticity of a short fibre reinforced composite with 20 % volume of glass fibre (E = 72.5 GPa) in a nylon matrix (E = 2.8 GPa) if the fibres are randomly oriented and have a length of 3.2 mm, and an aspect ratio of 320. (Ans: 8.9 GPa).

15.5 Glass fibres with a diameter of 0.03 mm are used to reinforce an epoxy resin. The fibres are aligned in the direction of the load and constitute 40 % by volume of the composite. The yield stress of the epoxy is 30 MPa and it (the epoxy) may be assumed to be rigid-perfectly plastic. If $E_f = 70$ GPa, $\sigma_{fu} = 1400$ MPa, $E_m = 3.5$ GPa, and $\tau_{mY} = 1/2\sigma_{mY}$, determine:
    (a) The critical fibre length. (Ans: 1.4 mm).
    (b) The fibre length necessary to attain full fibre stress at the fibre mid point if the required composite strength is: (i) 80 MPa, (ii) 210 MPa. (Ans: 0.031 mm; 0.09 mm).
    (c) The tensile strength of the composite if $l = 0.2\ l_c$; $l = 4\ l_c$, and $l = 60\ l_c$, where $l_c$ is the critical fibre length. (Ans: 74 MPa, 508 MPa, 578 MPa).
    (d) Plot a graph showing the variation of fibre stress and interfacial shear stress with fibre length for $0.5\ l_c < l < 6\ l_c$.

15.6 A large particle composite consists of a cobalt matrix (E = 200 GPa) and tungsten carbide (WC) particles (E = 700 GPa). Calculate the upper and lower bounds of the modulus of elasticity of the resulting composite if the volume fraction of the WC is 60 %. (Ans: 350 GPa; 500 GPa).

15.7 Calculate the fibre fracture strength required to produce an aligned short fibre composite with a longitudinal tensile strength of 1500 MPa. The fibres have a diameter of 0.012 mm and a length of 6.5 mm. The fibre-matrix bond strength is 95 MPa, and the stress in the matrix when the composite fails is 7 MPa. A fibre volume fraction of 40 % is desired. (Ans: 3.8 GPa).

15.8 Determine the fibre volume fraction required to produce an aligned short fibre composite with a longitudinal tensile strength of 710 MPa. The fibres have a diameter of 0.012 mm, a length of 0.6 mm, and an ultimate tensile strength of 3.6 GPa. The fibre matrix bond strength is 60 MPa, while the stress in the matrix at composite failure is 8 MPa. (Ans: 0.28).

# Further Reading

Agrawal, B. D. and Broutman, L. J. <u>Analysis and Performance of Fibre Composites,</u> John Willey and Sons, New York, 1980.

Hull, D. <u>An Introduction to Composite Materials,</u> Cambridge University Press, Cambridge, 1980.

Holister, G. S. and Thomas, C. <u>Fibre Reinforced Materials,</u> Elsevier, Amsterdam, 1966.

# CHAPTER SIXTEEN

# CORROSION AND DEGRADATION OF MATERIALS

## 16.1   INTRODUCTION

The term corrosion refers to the chemical/electrochemical interaction between a metal and its environment involving the removal of the metal through its conversion to the oxide or compound. Corrosion is an oxidation/reduction process.

## 16.2   TYPES OF CORROSION

There are two main types of corrosion: direct corrosion and electrochemical corrosion.

### 16.2.1   Direct Corrosion

This refers to the corrosion of a metal by dry oxygen or other dry gas. It is sometimes referred to as oxidation, scaling or dry corrosion. It involves the reaction of the metal with a component of its environment to form the oxide or compound. The reaction is accompanied by a release in energy, which acts as the driving force for more reaction:

$$M + {}^1/_2 O_2 \longrightarrow MO - \Delta G \qquad\qquad (16.1)$$

The oxide can only form when the metal and gas are in contact. The product of the reaction forms a layer on the metal. Further corrosion can only take place by diffusion through the oxide layer (figure 16.1). Moreover, the oxide layer must be able to conduct electrons from the metal/oxide interface to the oxide/gas interface.

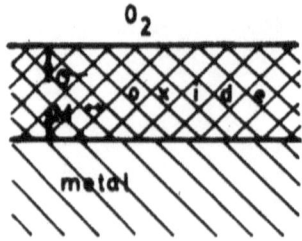

**Fig. 16.1** The mechanics of dry corrosion

The metal and oxygen form ions as (assuming a divalent metal):

$$M ---> M^{2+} + 2e^-$$ (16.2)

And at the oxide/oxygen layer:

$$^1/_2O_2 + 2e^- ---> O^{2-}$$ (16.3)

The ions then diffuse through the oxide layer and when they meet, react to form the oxide:

$$M^{2+} + O^{2-} ---> MO$$ (16.4)

From the above, it is clear that direct corrosion involves oxidation and reduction, that is to say it is a redox type reaction. The formation of metal ions represented by equation 16.2 involves release of electrons and hence is an oxidation, while the reaction represented by equation 16.3 is a reduction. If the oxide layer prohibits the diffussion of ions or does not conduct electrons, further oxidation is curtailed and hence the layer provides protection to the metal from further oxidation. This is the case with aluminium oxide and explains the corrosion resistence of aluminium. Alternatively, the oxide layer may crack or flake. In this case, the metal and gas are in constant contact. As a result, corrosion will proceed at a constant rate. The prevailing situation for a specific metal/oxide combination is determined by the Pilling-Bedworth (P-B) ratio which may be defined as:

$$P\text{-}B = \frac{A_o \rho_m}{A_m \rho_o}$$ (16.5)

where $A_o$ is the molecular weight of the oxide, $A_m$ is the molecular weight of the metal while $\rho_o$ and $\rho_m$ are the respective densities of the oxide and metal. If the P-B ratio is less than one, the oxide is porous and the rate of oxidation decreases with increase in the thickness of the oxide layer. If P-B is greater than two, the layer cracks or flakes and the metal surface is continuously exposed to the gas. This type of layer provides no protection to the metal. A P-B value between one and two results in a non-porous, non-flacking and hence protective oxide layer.

Other factors that may influence the protective ability of the oxide layer include the degree of adhesion between the metal and the oxide, the relative values of the respective coefficients of thermal expansion, and the melting point of the oxide.

The rate of dry corrosion depends on the P-B ratio. For oxide layers that flake, there is a linear relationship between the weight gain per unit area, W, and time i.e:

$$W = K_1 t$$ (16.6)

where $K_1$ is a constant, and t is the time. This rate law is applicable to the corrosion of sodium, potassium, and tantalum. For non-porous oxide layers that are adherent to the metal, a parabolic rate relationship results, i.e:

$$W = K_2t + K_3t^2 \qquad (16.7)$$

where $K_2$ and $K_3$ are other constants. This rate equation is noted for dry corrosion of iron, copper and cobolt among others. For P-B values between one and two, such as those applicale to aluminium, the corrosion rate is logarithmic, i.e:

$$W = K_4\log (K_5t + K_6) \qquad (16.8)$$

where the $K_i$'s are constants. These corrosion rates are shown schematically in figure 16.2.

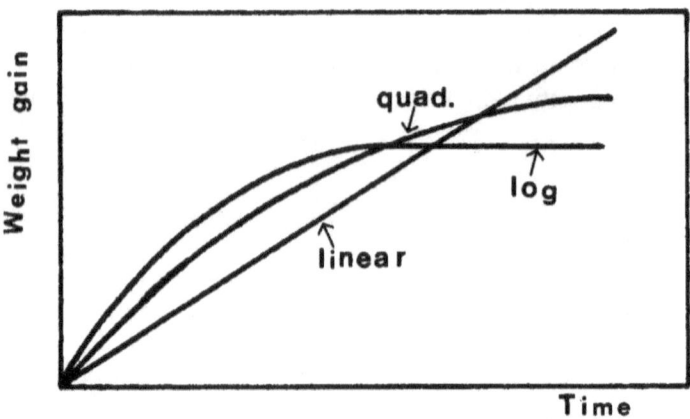

**Fig. 16.2** Schematic representation of rates of dry corrosion

The diffusion requires lattice defects and the diffusion rate depends on the size of these lattice defects, relative to the size and polarity of the ions. Since an increase in temperature causes an increase in rate of diffusion, the rate of direct corrosion increases with increase in temperature.

## 16.2.2   Electrochemical Corrosion

This refers to the corrosion that takes place in the presence of water or aqueous solutions. Its rate depends on the concentration of ions in the solution. Now, it will be remembered from school chemistry that water ionizes to form hydrogen and hydroxyl ions as:

$$H_2O ===== H^+ + OH^- \qquad\qquad (16.9)$$

The concentration of these ions: $[H^+] = [OH^-] = 10^{-7}$ kilomole/$m^3$ in pure water. Moreover, by the laws of chemical equilibrium, the product of the concentrations must always be constant, i.e:

$$[H^+] \times [OH^-] = \text{Constant} = 10^{-14} \qquad\qquad (16.10)$$

The alkalinity/acidity of the solution is expressed as the concentration of hydrogen ions. This is expressed as the pH, x, where:

$$[H^+] = 10^{-x} \text{ kmole}/m^3 \qquad\qquad (16.11)$$

A neutral solution has a pH value of 7. The solution is acidic if $x < 7$ and alkaline if $x > 7$.

In electrochemical corrosion, the chemical process is accompanied by a flow of electric current. Thus the corrosion system is divided into cathodic and anodic areas. This sets up what is known as a corrosion or galvanic cell. Indeed, electrochemical corrosion is best illustrated by a corrosion cell in which two metals (e.g., zinc and copper) are placed together in a copper sulphate solution (figure 16.3). Corrosion of the zinc electrode (i.e., zinc going into solution) starts as soon as the two metals are connected by a conducting path. Simultaneously, reduction of copper (i.e., deposition of copper onto the electrode) occurs.

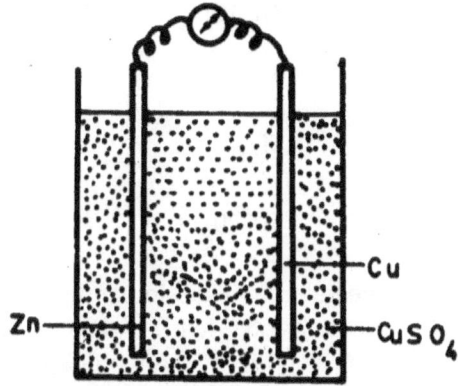

**Fig. 16.3** A corrosion or galvanic cell

The following conditions must therefore be met for electrochemical corrosion to take place:

(i) There must be two areas with differing potential (the anodic and cathodic areas).

(ii) The two areas must be in the same aqueous solution.

(iii) The two areas must be connected by an electric conducting path.

At both the anodic and cathodic areas, chemical reactions take place. The reactions that take place at the anode are termed anodic reactions. They are all oxidation reactions and hence release electrons. Some common anodic reactions are:

1. Metal going into solution

$$M \longrightarrow M^{2+} + 2e^- \qquad (16.12)$$

2. Formation of an oxide

$$M + 2OH^- \longrightarrow MO + H_2O + 2e^- \qquad (16.13)$$

3. Formation of a hydroxide

$$M + 2OH^- \longrightarrow M(OH)_2 + 2e^- \qquad (16.14)$$

4. Formation of a compound of the metal

$$M + 2X^- \longrightarrow MX_2 + 2e^- \qquad (16.15)$$

The electrons travel through the conducting path (constituting the electric current) and are neutralized at the cathode in "cathodic reactions". All cathodic reactions are therefore reduction reactions and include:

1. Deposition of a metal:

$$M^{2+} + 2e^- \longrightarrow M \qquad (16.16)$$

2. Evolution of hydrogen:

$$2H^+ + 2e^- \longrightarrow H_2 \qquad (16.17)$$

3. Change (decrease) in anion valency:

$$2M^{3+} + 2e^- \longrightarrow 2M^{2+} \qquad (16.18)$$

4. Change (increase) in cation valency:

$$2X^{2-} + 2e^- \longrightarrow 2X^{3-} \qquad (16.19)$$

5. Formation of hydroxyl ions:

$$1/2 O_2 + H_2O + 2e^- \longrightarrow 2OH^- \qquad (16.20)$$

It follows from the above that whenever two dissimilar metals are placed in the same solution an electromotive force (emf) is developed. If each metal is placed in a molar solution of its ions, the resulting cell is termed an electrochemical cell and a standard emf will be developed. Each metal immersed in its molar solution at 25 °C constitutes a standard half cell. The magnitude of the emf is proportional to the driving force for the redox reaction. The direction of the force depends on the particular metals.

So the metals can be arranged in a series termed electrochemical series in which any metal is anodic with respect to all those below it and cathodic with respect to all metals above it. The reference electrode is defined as a platinum metal immersed in a molar solution of hydrogen ions. The potentials developed by each metal when immersed in a molar solution of its ions and connected by a conducting path to the reference electrode is termed the standard potential, V°, of the metal concerned. Part of the electrochemical series with the standard potentials is given in Table 16.1.

**Table 16.1**   Part of the electrochemical series

| Li | -3.04 V | Cr | -0.74 V |
|----|---------|----|---------|
| K | -2.92 V | Fe | -0.44 V |
| Na | -2.71 V | $H_2$ | 0.00 V |
| Mg | -2.37 V | Cu | +0.34 V |
| Al | -1.66 V | Ag | +0.80 V |
| Zn | -0.76 V | Au | +1.50 V |

Going back to the standard electrochemical cell, the total potential difference developed is the algebraic sum of the two standard potentials. For example if metal $M_1$ has a standard potential of $-V°_1$, and metal $M_2$ has the standard potential $+V°_2$, the potential difference developed when $M_1$ and $M_2$ form an electrochemical cell is $\Delta V° = V°_2 - V°_1$. For the reaction to be sponteneous, $\Delta V°$ must be positive. The above applies only to standard electrodes i.e., pure metals in their molar solutions at 25 °C. The potential difference developed will be affected by the temperature, concentration of the solution, etc. For non-standard conditions, the developed potential is given by:

$$\Delta V = \left[V_2^O - V_1^O\right] - \frac{RT}{nF} \ln \left| \frac{M_1^{n+}}{M_2^{n+}} \right| \qquad (16.21)$$

where R is the gas constant, F is Faraday's constant and n is the valency.

Electrochemical corrosion cannot take place without current flowing in the external circuit. It can therefore be decelerated, accelerated, stopped or reversed by applying an external potential difference. An electrode whose standard electrode potential has been shifted from its standard value by applying an opposing potential is said to be "polarized".

When metal ions go into solution at the anode in electrochemical corrosion, the compounds formed may either stick to the electrode or form a precipitate. In the cases where a precipitate is formed, the corrosion products do not protect the electrode from further corrosion. The rate of corrosion is then controlled by how fast the cathode can absorb electrons. The reaction is said to be under cathodic control. When the reaction produces adherent products, some protection is provided to the electrode and the rate of reaction (corrosion) is controlled by how fast diffusion through the protective layer can take place. Thus it is the events happening at the anode, which control the rate of reaction and the corrosion is said to be under anodic control. If the adherent corrosion products prevent further corrosion altogether, the electrode is said to have been rendered passive or immune.

The rate of electrochemical corrosion may be determined by defining the corrosion penetration rate, CPR which is the loss in thickness per unit time. This may be expressed as:

$$CPR = \frac{KW}{\rho At} \qquad (16.22)$$

where W is the weight loss, K is a constant, A is the exposed area while $\rho$ is the density of the metal. Alternatively, it may be expressed in terms of the current density associated with the reaction, i.e:

$$\text{Corrosion rate} = \frac{i}{nF} \qquad (16.23)$$

## 16.3   POURBAIX DIAGRAMS

From the foregoing, the rate of electrochemical corrosion seems to depend on a lot of factors: the concentration and nature of the solution, the nature of the corrosion products, the solution pH, the potential of the two electrodes, the quantity and direction of the applied external potential difference (if any), etc. This information can be summarized in an equilibrium diagram of electrochemical reaction for any metal/solution system. Such a diagram is termed a Pourbaix diagram. A simplified Pourbaix diagram for the iron water system is shown in figure 16.4. Under the conditions marked "corrosion", corrosion takes place with the corrosion products being as shown in the figure. In the region marked immunity, no corrosion takes place due to low concentration of the $Fe^{2+}$ ions. In the passivity regions, iron is rendered passive as an electrode due to the formation of an adherent layer of $Fe_2O_3$.

As explained hereafter, Pourbaix diagrams may help in the control or prevention of corrosion.

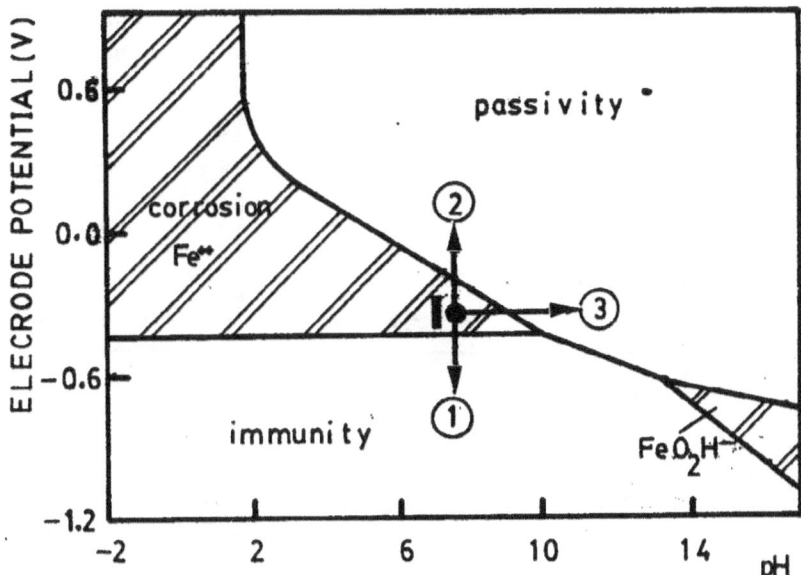

**Fig. 16.4** A simplified Pourbaix diagram for the iron-water system

## 16.4   FORMS OF CORROSION

Corrosion may also be classified based on observations of corrsion phenomenon in real life. Using this method, nine forms of corrosion may be identified. These include:

1. Uniform Attack. This is corrosion where the intensity is equal over the entire surface of the component. An example of this is the general tarnishing of silverware. This is basically similar to dry corrosion.

2. Galvanic Corrosion. Refers to corrosion that takes place when two dissimilar metals are electrically connected and in contact with the same solution. Real life examples are screws of one material used to hold together pieces of a different material, or water tubes held in place by supports made from a different material.

3. Crevice Corrosion. This refers to localized corrosion due to concentration differences at two positions of the same metal. This setup is termed a concentration cell. This form of corrosion is prevalent around crevices such as the gap between riveted plates as illustrated in figure 16.5. The water inside the gap is stagnant. As a result, it becomes depleted in oxygen. This

renders the surrounding parts anodic with respect to the parts of the metal exposed to the atmosphere. The metal at the crevice therefore goes into solution releasing electrons in the process. The electrons then travel through the metal to the cathodic part where the cathodic reaction involving formation of hydroxyl ions as represented by equation 16.20 take place.

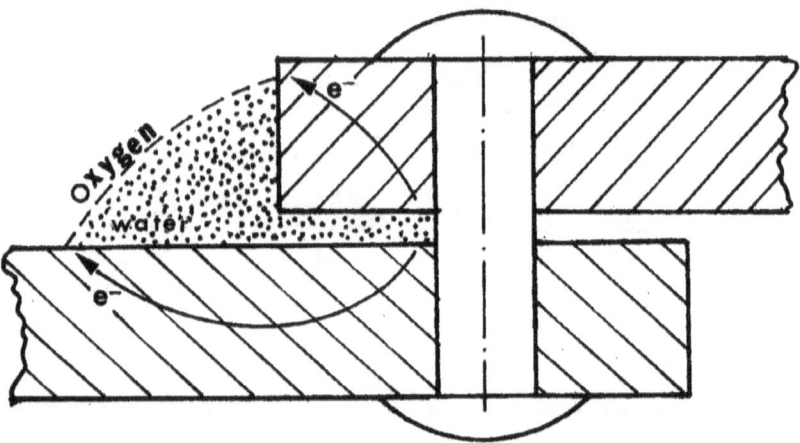

**Fig. 16.5** Crevice corrosion

<u>4. Pitting.</u> The mechanisms of pitting are the same as those for crevice corrosion but result from a stagnant pool of water on a surface. Again, the concentration of oxygen is higher at the edges of the drop compared to the centre of the drop. A concentration cell is set up causing corrosion at the centre and creating a pit (figure 16.6). Other sources of inhomogeneity that may lead to pitting include differences in concentration of a solution in contact with a metal, breaks in the protective layer, inherent inhomogeneity due to variations in chemical composition, and differences in stress such as those caused by residual stress.

<u>5. Intergranular Corrosion.</u> This refers to corrosion along the grain boundaries and usually takes place in stainless steels. In this particular case, $Cr_{23}C_6$ forms at the grain boundaries when the material is heated to a temperature between 500 ºC and 800 ºC. This leads to the formation of a chromium free zone around the grain boundary, which renders them liable to corrosion. This is the cause of weld decay considered in chapter 9. Intergranular corrosion causes areas of weakness in a component that may otherwise look normal.

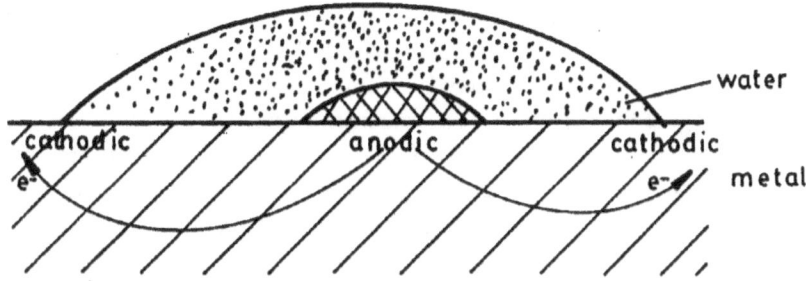

**Fig. 16.6** Pitting

6. Selective leaching. This takes place in alloys consisting of two or more metals. One of the elements forming the alloy is oxidized preferentially leaving a porous mass. Such corrosion is particularly noticeable in brass. Zinc is preferentially oxidized leaving behind porous cooper structure. This explains the tendency of brass battery terminals to turn red with age.

7. Erosion. This is a combined chemical/mechanical effect especially observed where fluid motion is involved e.g., water pipes, propellers, turbine blades, valves, etc.

8. Stress Corrosion. Refers to the combined effect of a static stress and a corrosive environment. In these circumstances, cracks will develop and grow perpendicular to the maximum principal stress at stresses much lower than those required to propagate the crack in an inert environment. The effect is very specific to certain combinations of metal and environment. Examples of combinations susceptible to stress corrosion are stainless steel in an environment containing chloride ions, brass in an ammonia environment, etc.

9. Corrosion Fatigue. This is similar to stress corrosion except that the stress is continuously varying. In other words, it is the combined effect of fatigue and a corrosive environment.

## 16.5 PREVENTION OF CORROSION

The point marked I in figure 16.4 represents the conditions when iron (electromotive force = -0.44V) is placed in pure water (pH = 7). The point is in the "corrosion" region and hence under normal conditions, corrosion will take place. The 3 arrows show the 3 ways in which we can move from this point to the non-corrosion regions:

1. Connecting a metal which is anodic with respect to iron e.g., zinc, to iron in the same solution. This causes the net potential to be lowered into the

"immunity" region (shown by arrow 1). This method is termed cathodic protection. The more anodic metal is corroded in place of iron i.e., it is used sacrificially. The same effect may be achieved by applying an external DC voltage to move the conditions into the immunity region.

Cathodic protection is applied in several industrial settings. Boiler tubes, for example, may be protected against corrosion by attaching plates of zinc inside the boiler. Applying an external DC voltage to the line will protect buried pipelines.

2. The external potential may also be applied to move the conditions into the "passivity" region. This forces the formation of the protective layer, which greatly slows down the rate of corrosion. This method of corrosion prevention (shown in figure 16.4 by arrow 2) is termed anodic protection and again may be used for buried pipelines.

3. Increasing the alkalinity (pH) of the solution moves the conditions in the direction shown by arrow 3. This is termed corrosion control by control of pH. The treated water fed into boilers has NaOH added to it to bring the pH to the range 11-12 to prevent corrosion of boiler tubes.

Other methods that may be used to prevent corrosion include:

(i) The use of inhibitors e.g., chromates and phosphates for iron.

(ii) Choice of a metal with inherent corrosion resistant properties e.g., stainless steels or aluminium.

(iii) Avoiding the direct contact of two dissimilar metals hence avoiding the creation of a corrosion cell. This may be done, for example, by using plastic washers to separate the two metals.

(iv) Use of protective coatings to ensure that the metal and environment do not come into contact. The protective coatings may be paints, vanishes, lacquers, metallic coatings, oil, grease, etc.

## 16.6   DEGRADATION OF POLYMERS

Degradation refers to the deterioration of mechanical properties of polymers as a result of the interuction with the environment. It is synonimous with corrosion in metals. Unlike corrosion that is electrochemical in nature, polymer degradation is mainly physiochemical. The various mechanisms responsible for it include swelling, dissolution and bond rupture.

Swelling results when the polymer is in contact with liquids. The liquid diffusses into the spaces between the chains forcing them to move apart. The increased molecular separation results in weaker secondary bonds. Some liquid hydrocarbons may also physically dissolve polymers. This effect decreases with increasing molecular weight, increased cross linking and higher degree of crystallinity. The resistance of polymers to inorganic acids and alkalis is however much better.

Bond rupture, also termed scission, involves the break up of the polymer chains. The result of this is reduced molecular weight, and hence deterioration of physical and mechanical properties. Bond rupture may be caused by radiation, heat or chemicals. Radiations that may cause bond rupture include electron beams, X-rays, γ-rays, β-rays, and ultra violet radiation. These penetrate the material and interrupt in various ways with the atoms and their electrons. For example, the radiation may knock an electron off its orbit thus leading to ionization. This in turn may lead to rupture of a primary bond or breakage of a cross-link.

The main causes of chemical scission in polymers are oxygen and ozone. Their effect is particularly pronounced if the polymer has unused double bonds. A good example is the effect of ozone on rubber:

$$--R--C==C--C--R' + O_3 \longrightarrow --R--C==O + O=C--R' + O \quad (16.4)$$
$$\qquad | \quad | \qquad\qquad\qquad\qquad | \qquad\qquad |$$
$$\qquad H \quad H \qquad\qquad\qquad\qquad H \qquad\qquad H$$

Scission may also occur due to high temperature. The elevated temperature may cause chemical reactions resulting in the release of gases. The end result is a loss in weight. The severity of the effect depends on the bond strength.

Weathering, the deteriation of properties in a polymer subjected to adverse weather conditions, is a mixture of two or more of the above mechanisms. The most severe is oxidation following exposure to ultraviolet radiation.

## PROBLEMS

16.1 Give three examples of anodic reactions and three of cathodic reactions that can occur in electrochemical corrosion.

16.2 Explain what is meant by the following terms: (i) stress corrosion, (ii) corrosion fatigue, (iii) fretting.

16.3 Compare and contrast the principles involved in cathodic and anodic protection of a metal against corrosion. Using a simplified Pourbaix diagram, show how the two methods can be applied for the iron-water system.

State the other methods that can be used for protection of iron (or steel) against corrosion.

16.4 Give an account of how direct corrosion of a metal takes place.

16.5 Explain how pitting may take place on a steel surface exposed to the atmosphere.

16.6 Describe a method (explaining the principle involved) that can be used for protecting (i) a buried pipeline, (ii) steel boiler tubes; (iii) an exposed steel structure, against corrosion.

16.7 (a) Describe the mechanism of rust formation beneath a drop of water on a steel surface.

(b) Explain fully the meanings of the following terms: (i) corrosion fatigue and its effects on the S-N curve in steels (ii) anodizing (iii) electroplating.

16.8 Using a sketch of a simple galvanic cell explain what takes place during electrolytic corrosion of a metal.

16.9 Give an account of how direct corrosion of a metal takes place and explain why you would expect the rate (of direct corrosion) to increase with increase in temperature.

16.10 What are the two main types of corrosion and under what conditions does each occur?

16.11 Describe a (simplified) Pourbaix diagram for the iron water system and explain how it is used to show the conditions under which a particular method of corrosion protection applies.

## Further Reading

Uhlig, H.H. and Revie, R.W. <u>Corrosion and Corrosion Control,</u> 3rd ed. John Wiley and Sons, New York, 1985.

Fortana, M. G. <u>Corrosion Engineering</u>, 3rd ed. McGraw--Hill, New York, 1986.

Evans, U.R. <u>An Introduction to Metallic Corrosion,</u> 3rd ed. Edward Arnold, Baltimore, 1981.

# APPENDIX

## PERIODIC TABLE OF THE ELEMENTS

| IA | IIA | IIIB | IVB | VB | VIB | VIIB | VIII | VIII | VIII | IB | IIB | IIIA | IVA | VA | VIA | VIIA | O |
|----|-----|------|-----|-----|------|------|------|------|------|-----|-----|------|-----|-----|-----|------|---|
| 1<br>H<br>1.008 | | | | | | | | | | | | | | | | | 2<br>He<br>4.003 |
| 3<br>Li<br>6.939 | 4<br>Be<br>9.012 | | | | | | | | | | | 5<br>B<br>10.81 | 6<br>C<br>12.01 | 7<br>N<br>14.01 | 8<br>O<br>16.00 | 9<br>F<br>19.00 | 10<br>Ne<br>20.18 |
| 11<br>Na<br>22.99 | 12<br>Mg<br>24.31 | | | | | | | | | | | 13<br>Al<br>26.98 | 14<br>Si<br>28.09 | 15<br>P<br>30.97 | 16<br>S<br>32.06 | 17<br>Cl<br>35.45 | 18<br>Ar<br>39.95 |
| 19<br>K<br>39.10 | 20<br>Ca<br>40.08 | 21<br>Sc<br>44.96 | 22<br>Ti<br>47.90 | 23<br>V<br>50.94 | 24<br>Cr<br>52.00 | 25<br>Mn<br>54.94 | 26<br>Fe<br>55.85 | 27<br>Co<br>58.93 | 28<br>Ni<br>58.71 | 29<br>Cu<br>63.54 | 30<br>Zn<br>65.37 | 31<br>Ga<br>69.72 | 32<br>Ge<br>72.59 | 33<br>As<br>74.92 | 34<br>Se<br>78.96 | 35<br>Br<br>79.91 | 36<br>Kr<br>83.80 |
| 37<br>Rb<br>85.47 | 38<br>Sr<br>87.62 | 39<br>Y<br>88.91 | 40<br>Zr<br>91.22 | 41<br>Nb<br>92.91 | 42<br>Mo<br>95.94 | 43<br>Tc<br>(99) | 44<br>Ru<br>101.1 | 45<br>Rh<br>102.9 | 46<br>Pd<br>106.4 | 47<br>Ag<br>107.9 | 48<br>Cd<br>112.4 | 49<br>In<br>114.8 | 50<br>Sn<br>118.7 | 51<br>Sb<br>121.8 | 52<br>Te<br>127.6 | 53<br>I<br>126.9 | 54<br>Xe<br>131.3 |
| 55<br>Cs<br>132.9 | 56<br>Ba<br>137.3 | Rare<br>earth<br>series | 72<br>Hf<br>178.5 | 73<br>Ta<br>181.0 | 74<br>W<br>183.9 | 75<br>Re<br>186.2 | 76<br>Os<br>190.2 | 77<br>Ir<br>192.2 | 78<br>Pt<br>195.1 | 79<br>Au<br>197.0 | 80<br>Hg<br>200.6 | 81<br>Tl<br>204.4 | 82<br>Pb<br>207.2 | 83<br>Bi<br>209.0 | 84<br>Po<br>(210) | 85<br>At<br>(210) | 86<br>Rn<br>(222) |
| 87<br>Fr<br>(223) | 88<br>Ra<br>(226) | Acti-<br>nide<br>series | | | | | | | | | | | | | | | |

| Rare earth series | 57<br>La<br>138.9 | 58<br>Ce<br>140.1 | 59<br>Pr<br>140.9 | 60<br>Nd<br>144.2 | 61<br>Pm<br>(145) | 62<br>Sm<br>150.4 | 63<br>Eu<br>152.0 | 64<br>Gd<br>157.3 | 65<br>Tb<br>159.0 | 66<br>Dy<br>162.5 | 67<br>Ho<br>164.9 | 68<br>Er<br>167.3 | 69<br>Tm<br>168.9 | 70<br>Yb<br>173.0 | 71<br>Lu<br>175.0 |
|---|---|---|---|---|---|---|---|---|---|---|---|---|---|---|---|
| Actinide series | 89<br>Ac<br>(227) | 90<br>Th<br>232.0 | 91<br>Pa<br>(231) | 92<br>U<br>238.0 | 93<br>Np<br>(237) | 94<br>Pu<br>(242) | 95<br>Am<br>(243) | 96<br>Cm<br>(247) | 97<br>Bk<br>(247) | 98<br>Cf<br>(249) | 99<br>Es<br>(254) | 100<br>Fm<br>(253) | 101<br>Md<br>(256) | 102<br>No<br>(254) | 103<br>Lw<br>(257) |

KEY: The top figure denotes the atomic number; the middle characters the symbol; while the lower figure denotes the atomic mass. Mass numbers in brackets denote artificially produced elements.

# INDEX

true, 50, 53, 54
Striations
Substitutional
    impurity, 245
    solid solution, 70
Supesaturated solution, 106,
    108, 111, 127

T
Tempered martensite, 133, 135
Tensile testing, 48, 53, 63, 150,
    266, 273
Tertiary creep, 176
Tetragonal crystal system, 14,
    15, 109
Thermit welding, 103
Thermoplastics, 213, 214, 215,
    227, 280
Thermosets, 213, 215, 227, 280
Toughness, 53, 63, 98, 112, 191
TTT curves, 129, 130, 131, 137
Titanium, 17, 19, 23
Titanium alloys, 121, 122, 123,
    124
Twin boundaries, 41

U
Unit cells, 14, 17, 18, 19, 20, 26,
    69
Urwin's constants, 49

V
Vacancies, 33, 34, 35, 176, 178,
    243
Valence electrons, 6
Van der Waals bonding, 12
Vickers hardness test, 59, 60, 61
Viscoelasticity, 216, 218, 222
Viscosity, 223, 254
Void coalescence, 191, 192

Voigt—Kelvin model, 225, 226,
    227
Vulcanization, 216, 231

W
Weight fraction, 262, 263
Weld decay, 140
White cast iron, 96, 97
Widmanstatten structure, 92
Wrought alloys, 104, 105, 112

X
X-ray diffraction, 28, 29

Y
Yield stress, 48
Young's modulus. *See* Modulus
    of elasticity